W9-AMM-769

PARCC REHEARSAL AND SKILLS MASTERY

This Book Includes:

- **Access to Online Practice Assessments**
  - 2 PARCC Rehearsal Practice Tests
  - Self-paced learning and personalized score reports
  - Strategies for building speed and accuracy
  - Instant feedback after completion of the Assessments
- **Standards based practice**
  - Operations and Algebraic Thinking
  - Number & Operations in Base Ten
  - Number & Operations - Fractions
  - Measurement and Data
  - Geometry
- **Detailed answer explanations for every question**

Complement Classroom Learning All Year

Using the Lumos Study Program, parents and teachers can reinforce the classroom learning experience for children. It creates a collaborative learning platform for students, teachers and parents.

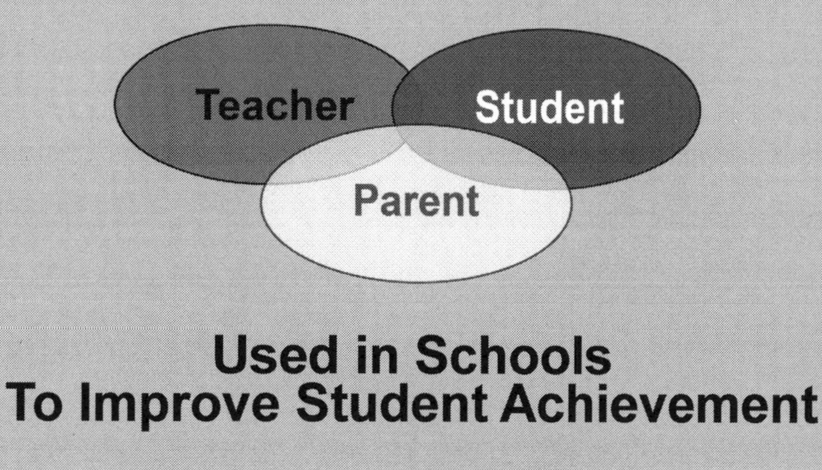

Used in Schools To Improve Student Achievement

Lumos Learning
Developed by Expert Teachers

PARCC Test Prep: 3rd Grade Math Practice Workbook and Full-length Online Assessments: PARCC Study Guide

Contributing Editor - Keyana M. Martinez
Contributing Editor - LaSina McLain-Jackson
Contributing Editor - Greg Applegate
Executive Producer - Mukunda Krishnaswamy
Designer - Mirona Jova
Database Administrator - R. Raghavendra Rao

COPYRIGHT ©2018 by Lumos Information Services, LLC. **ALL RIGHTS RESERVED.** No part of this work covered by the copyright hereon may be reproduced or used in any form or by means graphic, electronic, or mechanical, including photocopying, recording, taping, Web distribution or information storage and retrieval systems- without the written permission of the publisher.

NGA Center/CCSSO are the sole owners and developers of the Common Core State Standards, which does not sponsor or endorse this product. © Copyright 2010. National Governors Association Center for Best Practices and Council of Chief State School Officers.

PARRC ® is a registered trademark of Parcc, Inc., which does not sponsor or endorse this product.

ISBN-10: 1-946795-28-3

ISBN-13: 978-1-946795-28-1

Printed in the United States of America

For permissions and additional information contact us

Lumos Information Services, LLC
PO Box 1575, Piscataway, NJ 08855-1575
http://www.LumosLearning.com

Email: support@lumoslearning.com
Tel: (732) 384-0146
Fax: (866) 283-6471

Table of Contents

Introduction

This book is designed to help students get Partnership for Assessment of Readiness for College and Careers (PARCC) rehearsal along with standards aligned rigorous skills practice. Unlike a traditional book, this Lumos tedBook offers two full-length practice tests online. Taking these tests will not only help students get a comprehensive review of standards assessed on the PARCC, but also become familiar with the question types.

After students take the test online, educators can use the score report to assign specific lessons provided in this book.

Students will obtain a better understanding of each standard and improve on their weaknesses by practicing the content of this workbook. The lessons contain rigorous questions aligned to the state standards and substandards. Taking the time to work through the activities will afford students the ability to become proficient in each grade level standard.

Quick facts about the PARCC Test

- PARCC is based on Common core state standards.
- Required for all students in grades 3-8.
- Each student will be assessed in ELA and Math.
- Between 3 to 4 ½ hours total testing time per grade.
- Students can opt for either Computer-based test or a paper and pencil test while students with hardship can take Partnership for Assessment of Readiness for College and Careers (PARCC) in place of the general education State tests in grades 3-8.
- The PARCC ELA is administered in 3 Units where Unit 1 (Reading), Unit 2 (Reading & Writing) and Unit 3 (Writing only). PARCC Math is administered in 4 units which are Unit 1, Unit 2, Unit 3, and Unit 4 for Grades 3 to 5 and 3 units for grade 6 to 8 which are Unit 1, Unit 2 and Unit 3.

Mathematics Estimated Time on Task in Minutes				
Grade	**Unit 1**	**Unit 2**	**Unit 3**	**Unit 4**
3	60	60	60	60
4	60	60	60	60
5	60	60	60	60
6	80	80	80	NA
7	80	80	80	NA
8	80	80	80	NA

How Can the Lumos Study Program Prepare Students for PARCC Tests?

At Lumos Learning, we believe that year-long learning and adequate practice before the actual test are the keys to success on these standardized tests. We have designed the Lumos study program to help students get plenty of realistic practice before the test and to promote year-long collaborative learning.

This is a Lumos tedBook™. It connects you to Online Workbooks and additional resources using a number of devices including Android phones, iPhones, tablets and personal computers. The Lumos StepUp Online Workbooks are designed to promote year-long learning. It is a simple program students can securely access using a computer or device with internet access. It consists of hundreds of grade appropriate questions, aligned to the new Common Core State Standards. Students will get instant feedback and can review their answers anytime. Each student's answers and progress can be reviewed by parents and educators to reinforce the learning experience.

© Lumos Information Services 2018 | LumosLearning.com

The StepUp program also gives you access to EdSearch and Coherence map.

Lumos EdSearch is a safe search engine specifically designed for teachers and students. Using EdSearch, you can easily find thousands of standards aligned learning resources such as questions, videos, lessons, worksheets and apps. Teachers can use EdSearch to create custom resource kits to perfectly match their lesson objective and assign them to one or more students in their classroom.
To access the EdSearch tool, use the search box after you log into Lumos StepUp.

The Lumos Standards Coherence map provides information about previous level, next level and related standards. It helps educators and students visually explore learning standards. It's an effective tool to help students progress through the learning objectives. Teachers can use this tool to develop their own pacing charts and lesson plans. Educators can also use the coherence map to get deep insights into why a student is struggling in a specific learning objective.

Teachers can access the Coherence maps after logging into the StepUp Teacher Portal or by directly visiting http://www.lumoslearning.com/a/coherence-map

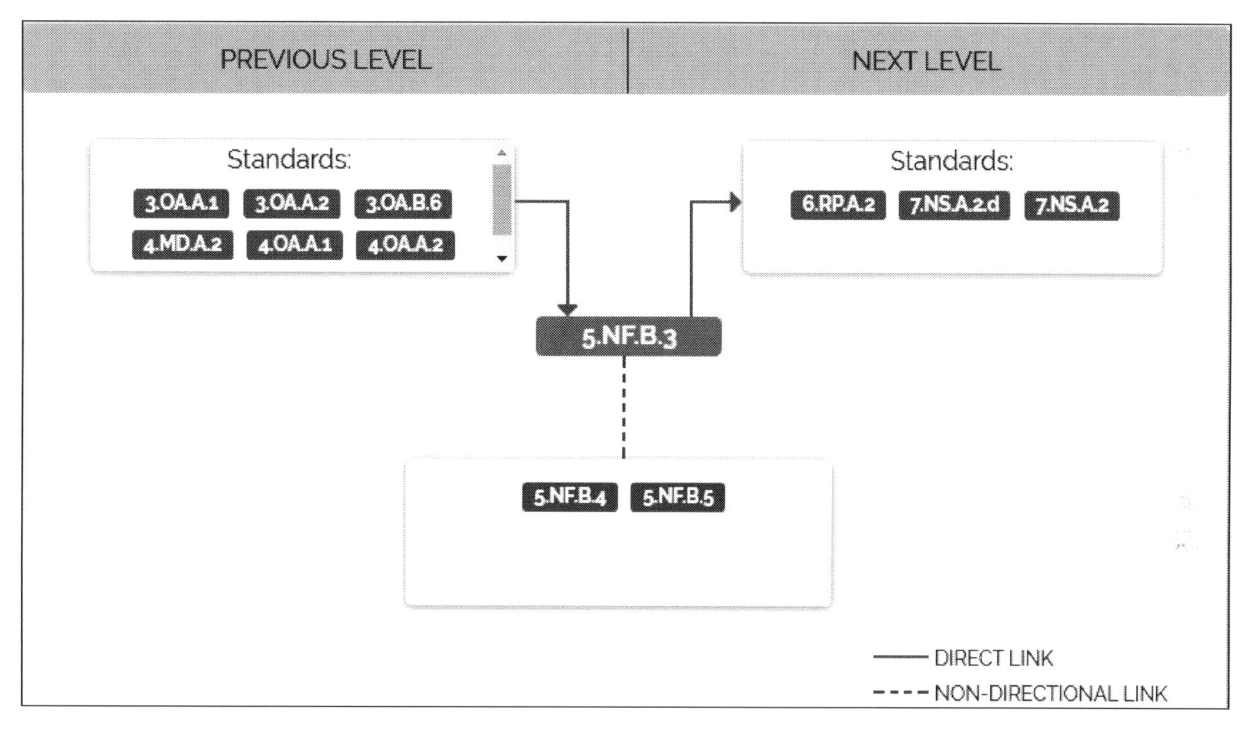

How to access the Lumos PARCC Online Assessments

***Important Instruction*: Please note that Lumos PARCC practice tests are provided in the online format only. Use the instructions provided on this page to access two full-length assessments.**

First Time Access:

Using a personal computer with internet access:	Using a smartphone or tablet:
Go to http://www.lumoslearning.com/a/workbooks Select your State and enter the following access code in the Access Code field and press the 'Submit' button. Access Code: PARCCG3M-14692-P State Select State Access Code Please enter your Access Code Submit	Scan the QR Code below and follow the instructions.

In the next screen, click on the "Register" button to register your user name and password.

Login to your Account

If you are not registered, please Register.

Login:

Password:

LOGIN

Forgot password?

Subsequent Access:

After you establish your user id and password for subsequent access, simply login with your account information.

What if I buy more than one Lumos Study Program?

Please note that you can use all Online resources with one User ID and Password. If you buy more than one book, you will access them with the same account.

Go back to the **http://lumoslearning.com/a/workbooks** link, select your state and enter the access code provided in the second book. In the next screen simply login using your previously created account.

© Lumos Information Services 2018 | LumosLearning.com

How to use this book effectively

The Lumos Program is a flexible learning tool. It can be adapted to suit a student's skill level and the time available to practice before standardized tests. Here are some tips to help you use this book and the online workbooks effectively:

The Lumos Program is a flexible learning tool. It can be adapted to suit a student's skill level and the time available to practice before standardized tests. Here are some tips to help you use this book and the online resources effectively:

Students

- The standards in each book can be practiced in the order designed, or in the order of your own choosing.
- Complete all problems in each workbook.
- Take the first practice test online.
- Have open-ended questions evaluated by a teacher or parent, keeping in mind the scoring rubrics.
- Take the second practice test as you get close to the official test date.
- Complete the test in a quiet place, following the test guidelines. Practice tests provide you an opportunity to improve your test taking skills and to review topics included in the New York State Test.

Parents

- Familiarize yourself with the PARCC Test format and expectations.
- Help your child use Lumos StepUp® PARCC Test Online Assessments by following the instructions in "How to access the Lumos PARCC Online Assessments" section of this chapter.
- Review your child's performance in the "Lumos PARCC Online Assessments" periodically. You can do this by simply asking your child to log into the system online and selecting the subject area you wish to review.
- Get useful information about your school by downloading the Lumos SchoolUp™ app. Please follow directions provided in "How can I Download the App?" section of this chapter.

How to create a teacher account

- You can use the Lumos online programs along with this book to complement and extend your classroom instruction.
- Get a Teacher account by visiting **LumosLearning.com/a/parccg3m**

  This Lumos StepUp® Basic teacher account will help you:

  - Create up to 30 student accounts
  - Review the online work of your students
  - Get insightful student reports
  - Discover standards aligned videos, apps and books through EdSearch
  - Easily access standards
  - Create and share information about your classroom or school events
  - Use EdSearch and Coherence maps to create personalized learning paths for your students.

  NOTE: There is a limit of one grade and subject per teacher for the free account.

  Mobile Access to the Teacher Portal: To access your student reports on a mobile device, download the Lumos SchoolUp™ mobile app using the instructions provided in "How can I Download the App?" section of this chapter.

QR code for Teacher account

© Lumos Information Services 2018 | LumosLearning.com

Test Taking Tips

1) **The day before the test,** make sure you get a good night's sleep.

2) **On the day of the test,** be sure to eat a good hearty breakfast! Also, be sure to arrive at school on time.

3) **During the test:**

- **Read every question carefully.**
  - Do not spend too much time on any one question. Work steadily through all questions in the section.
  - Attempt all of the questions even if you are not sure of some answers.
  - If you run into a difficult question, eliminate as many choices as you can and then pick the best one from the remaining choices. Intelligent guessing will help you increase your score.
  - Also, make note of the question so that if you have extra time, you can return to it after you reach the end of the section. Try to erase the marks after you complete the work.
  - Some questions may refer to a graph, chart, or other kind of picture. Carefully review the graphic before answering the question.
  - Be sure to include explanations for your written responses and show all work.

- **While Answering Multiple-Choice (EBSR) questions.**
  - Completely fill in the bubble corresponding to your answer choice.
  - Read **all** of the answer choices, even if think you have found the correct answer.

- **While Answering TECR questions.**
  - Read the directions of each question. Some might ask you to drag something, others to select, and still others to highlight. Follow all instructions of the question (or questions if it is in multiple parts)

Name ______________________________ Date ________________

Chapter 1:
Operations and Algebraic Thinking

Lesson 1: Understanding Multiplication

You can scan the QR code given below or use the url to access additional EdSearch resources including videos and mobile apps related to *Understanding Multiplication*.

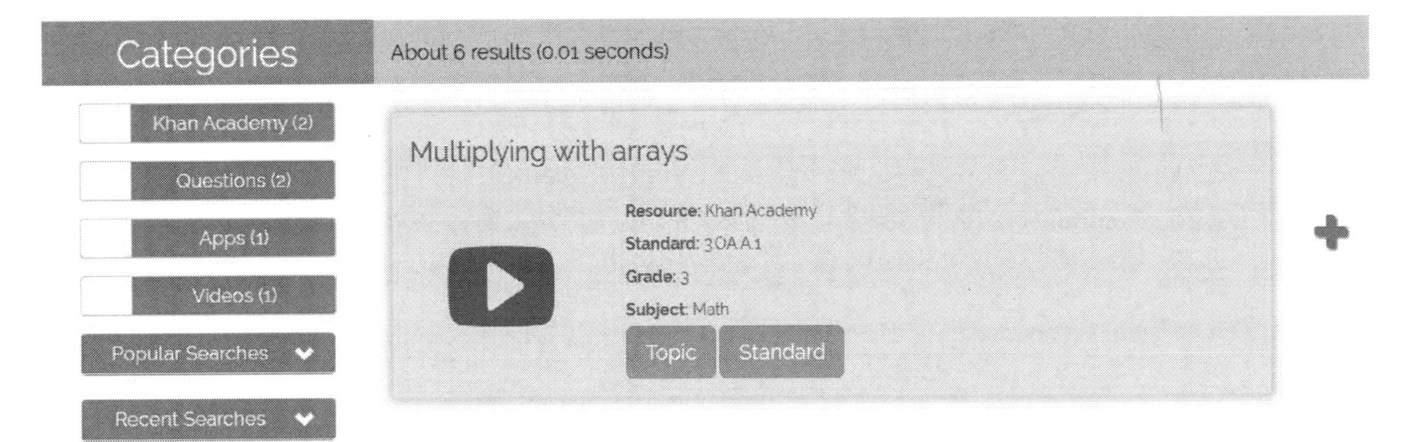

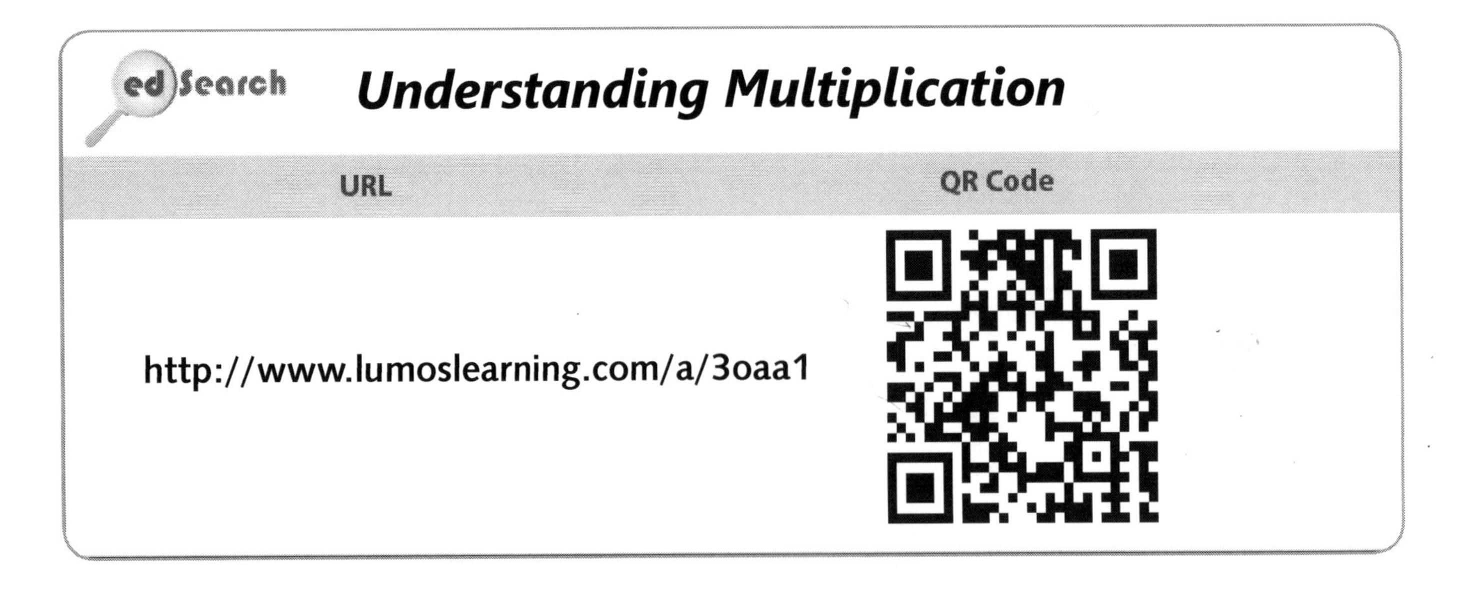

© Lumos Information Services 2018 | LumosLearning.com

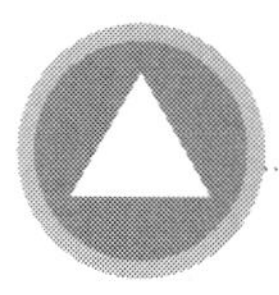

Name ______________________ Date ______________

1. **Which multiplication fact is being modeled below?**

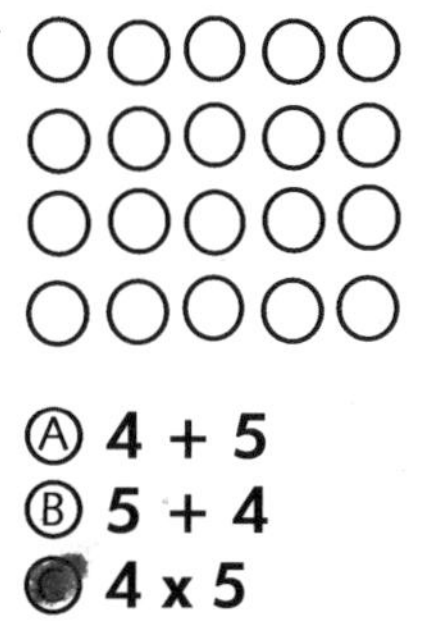

Ⓐ 3 x 10 = 30
Ⓑ 4 x 10 = 40
Ⓒ 4 x 9 = 36
Ⓓ 3 x 9 = 27

2. **Which numerical expression describes this array?**

○○○○○
○○○○○
○○○○○
○○○○○

Ⓐ 4 + 5
Ⓑ 5 + 4
Ⓒ 4 x 5
Ⓓ 4 x 4

3. **Which number sentence describes this array?**

Ⓐ 8 x 4 = 32
Ⓑ 7 + 5 = 12
Ⓒ 5 x 7 = 35
Ⓓ 7 x 4 = 28

© Lumos Information Services 2018 | LumosLearning.com

Name ______________________________ Date ________________

4. **Which number sentence describes this array?**

Ⓐ $2 \times 12 = 24$
Ⓑ $2 + 12 = 14$
Ⓒ $12 + 2 = 24$
Ⓓ $10 \times 2 = 20$

5. **Identify the multiplication sentence for the picture below:**

Ⓐ $4 \times 4 = 16$
Ⓑ $4 \times 3 = 12$
Ⓒ $3 \times 4 = 12$
Ⓓ $4 \times 2 = 8$

6. **What multiplication fact does this picture model?**

Ⓐ $4 \times 6 = 24$
Ⓑ $4 \times 7 = 28$
Ⓒ $6 \times 3 = 18$
Ⓓ $7 \times 4 = 28$

7. **Identify the multiplication sentence for the picture below:**

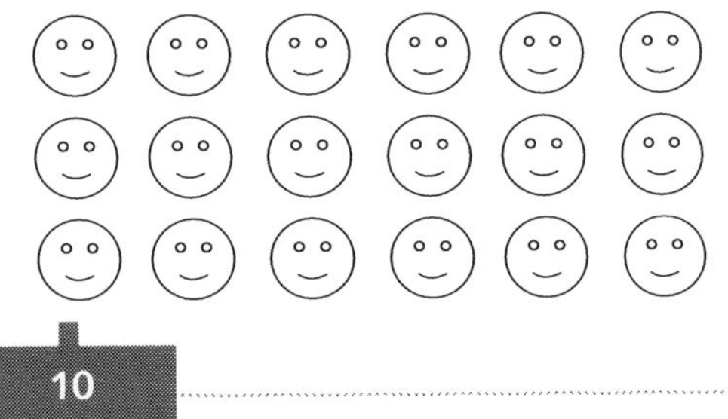

© Lumos Information Services 2018 | LumosLearning.com

Ⓐ 7 x 2 = 14
Ⓑ 7 x 3 = 21
Ⓒ 7 x 4 = 28
Ⓓ 6 x 3 = 18

8. Identify the multiplication sentence for the picture below:

Ⓐ 4 x 4 = 16
Ⓑ 3 x 6 = 18
Ⓒ 3 x 4 = 12
Ⓓ 3 x 5 = 15

9. Identify the multiplication sentence for the picture below:

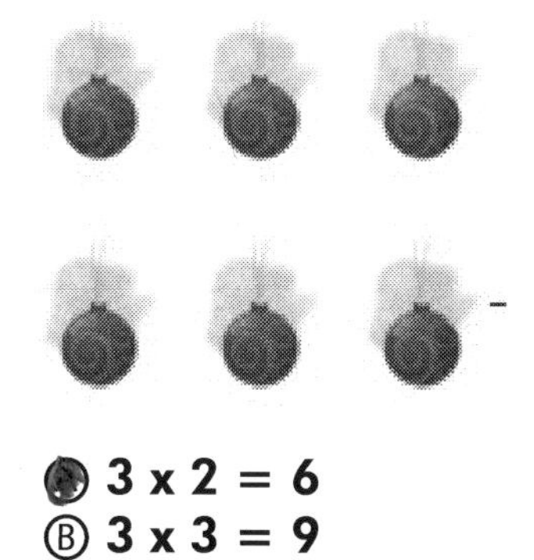

Ⓐ 3 x 2 = 6
Ⓑ 3 x 3 = 9
Ⓒ 4 x 2 = 8
Ⓓ 3 x 1 = 3

© Lumos Information Services 2018 | LumosLearning.com

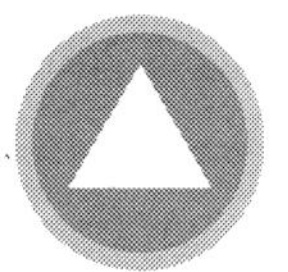

10. Identify the multiplication sentence for the picture below:

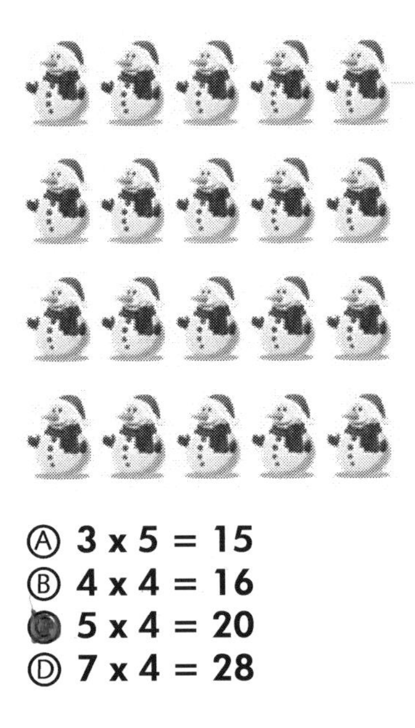

Ⓐ 3 x 5 = 15
Ⓑ 4 x 4 = 16
Ⓒ 5 x 4 = 20
Ⓓ 7 x 4 = 28

11. Identify the multiplication sentence for the picture below:

Ⓐ 2 x 5 = 10
Ⓑ 4 x 2 = 8
Ⓒ 4 x 1 = 4
Ⓓ 4 + 2 = 6

© Lumos Information Services 2018 | LumosLearning.com

12. Identify the multiplication sentence for the picture below:

Ⓐ 6 x 7 = 42
Ⓑ 6 x 8 = 48
Ⓒ 8 x 9 = 72
Ⓓ 8 x 8 = 64

13. Identify the multiplication sentence for the picture below:

Ⓐ 10 x 1 = 10
Ⓑ 9 x 2 = 18
Ⓒ 2 x 10 = 20
Ⓓ 5 x 4 = 20

14. Identify the multiplication sentence for the picture below:

© Lumos Information Services 2018 | LumosLearning.com

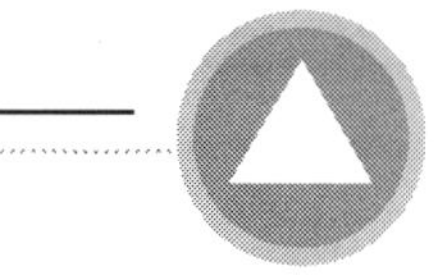

Ⓐ 5 x 5 = 25
Ⓑ 4 x 4 = 16
Ⓒ 4 x 6 = 24
Ⓓ 5 x 4 = 20

15. Identify the multiplication sentence for the picture below:

Ⓐ 3 x 1 = 3
Ⓑ 5 x 3 = 15
Ⓒ 3 x 2 = 6
Ⓓ 3 x 3 = 9

© Lumos Information Services 2018 | LumosLearning.com

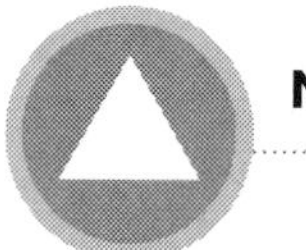

Name ______________________ Date ______________

Chapter 1

Lesson 2: Understanding Division

You can scan the QR code given below or use the url to access additional EdSearch resources including videos and mobile apps related to *Understanding Division*.

edSearch ***Understanding Division***

URL	QR Code
http://www.lumoslearning.com/a/3oaa2	

© Lumos Information Services 2018 | LumosLearning.com

Name ______________________ Date ______________

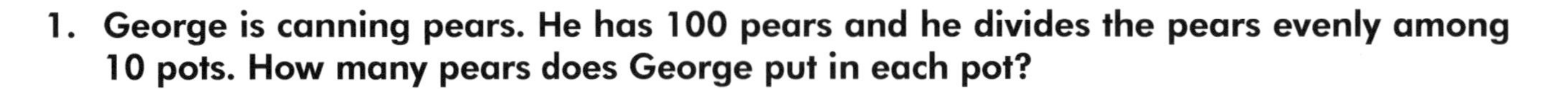

1. George is canning pears. He has 100 pears and he divides the pears evenly among 10 pots. How many pears does George put in each pot?

   Ⓐ 9 pears
   Ⓑ 5 pears
   Ⓒ 8 pears
   Ⓓ 10 pears

2. Marisa made 15 woolen dolls. She gave the same number of woolen dolls to 3 friends. How many dolls did Marisa give to each friend?

   Ⓐ 4 woolen dolls
   Ⓑ 3 woolen dolls
   Ⓒ 5 woolen dolls
   Ⓓ 6 woolen dolls

3. Lisa bought 50 mangos. She divided them equally into 5 basins. How many mangos did Lisa put in each basin?

   Ⓐ 10 mangos
   Ⓑ 8 mangos
   Ⓒ 5 mangos
   Ⓓ 7 mangos

4. Jennifer picked 30 oranges from the basket. If it takes 6 oranges to make a one liter jar of juice, how many one liter jars of juice can Jennifer make?

   Ⓐ 4 jars
   Ⓑ 3 jars
   Ⓒ 6 jars
   Ⓓ 5 jars

5. Miller bought 80 rolls of paper towels. If there are 10 rolls of paper towels in each pack, how many packs of paper towels did Miller buy?

   Ⓐ 6 packs
   Ⓑ 8 packs
   Ⓒ 7 packs
   Ⓓ 5 packs

© Lumos Information Services 2018 | LumosLearning.com

6. James takes 15 photographs of his school building. He gave the same number of photographs to 5 friends. How many photographs did James give to each friend?

Ⓐ 2 photographs
Ⓑ 3 photographs
Ⓒ 6 photographs
Ⓓ 5 photographs

7. Ron took 81 playing cards and arranged them into 9 equal piles. How many playing cards did Ron put in each pile?

Ⓐ 5 playing cards
Ⓑ 4 playing cards
Ⓒ 6 playing cards
Ⓓ 9 playing cards

8. Robert wants to buy 40 ice cream cups from the ice cream parlor. If there are 10 ice cream cups in each box, how many boxes of ice cream cups should Robert buy?

Ⓐ 6 boxes
Ⓑ 3 boxes
Ⓒ 4 boxes
Ⓓ 5 boxes

9. Marilyn wants to purchase 20 tiles. If the tiles come in packs of 5, how many packs should Marilyn buy?

Ⓐ 3 packs
Ⓑ 4 packs
Ⓒ 5 packs
Ⓓ 6 packs

10. There are 30 people running around the path. If the runners are evenly divided among the path's 5 lanes, how many people are running in each lane?

Ⓐ 6 runners
Ⓑ 5 runners
Ⓒ 8 runners
Ⓓ 4 runners

© Lumos Information Services 2018 | LumosLearning.com

Name ______________________ Date ______________

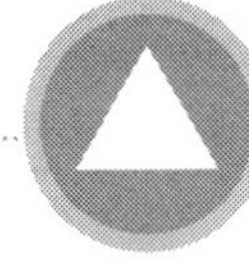

11. Sally is buying goodie bags for her class. She needs 24 bags in all. If the bags come in packs of 3. How many packs does Sally need?

Ⓐ 21 packs
Ⓑ 3 packs
Ⓒ 24 packs
Ⓓ 8 packs

12. Mr. Johnson is planting a garden. He wants to use all of his 44 seeds and wants to make 4 rows of vegetables. How many seeds should he plant in each row?

Ⓐ 22 seeds
Ⓑ 11 seeds
Ⓒ 4 seeds
Ⓓ 88 seeds

13. Destiny, Jimmy, and Marcy have 32 marbles all together. Tommy adds 4 marbles to the set. If the group of friends wants to evenly divide the marbles so that each person has the same amount, how many marbles should each person receive?

Ⓐ 4 marbles
Ⓑ 8 marbles
Ⓒ 9 marbles
Ⓓ 10 marbles

14. Mr. Baker earned $100 for five days of work. If he made the same amount each day, how much money did he make per day?

Ⓐ $15 per day
Ⓑ $20 per day
Ⓒ $25 per day
Ⓓ $30 per day

15. Seth and his brother have collected 26 seashells on the beach. If they want to share them equally, how many seashells will each of them receive?

Ⓐ 6 seashells
Ⓑ 9 seashells
Ⓒ 26 seashells
Ⓓ 13 seashells

© Lumos Information Services 2018 | LumosLearning.com

Name ______________________ Date ______________

Chapter 1

Lesson 3: Applying Multiplication & Division

You can scan the QR code given below or use the url to access additional EdSearch resources including videos and mobile apps related to *Applying Multiplication & Division*.

edSearch **_Applying Multiplication & Division_**

URL	QR Code
http://www.lumoslearning.com/a/3oaa3	

© Lumos Information Services 2018 | LumosLearning.com

Name ______________________________ Date _______________

1. 54 x 3 = ?
   The product in this number sentence is ________.

   Ⓐ 54
   Ⓑ 162
   Ⓒ 3
   Ⓓ 54 and 3

2. The Snack Shop has twice as many popcorn balls as they do cotton candy. If there are 30 popcorn balls, how many cotton candies are there?

   Ⓐ 7
   Ⓑ 450
   Ⓒ 30
   Ⓓ 15

3. Monica has 56 DVD's in her movie collection. This is 8 times as many as Sue has. How many DVDs does Sue have?

   Ⓐ 8
   Ⓑ 6
   Ⓒ 7
   Ⓓ 10

4. Jonathan can do 7 jumping jacks. Marcus can do 4 times as many as Jonathan. How many jumping jacks can Marcus do?

   Ⓐ 28
   Ⓑ 8
   Ⓒ 4
   Ⓓ 7

5. Darren has seen 4 movies this year. Marsha has seen 3 times as many movies as Darren. How many movies has Marsha seen?

   Ⓐ 7
   Ⓑ 3
   Ⓒ 4
   Ⓓ 12

© Lumos Information Services 2018 | LumosLearning.com

6. Sarah is planting a garden. She will plant 4 rows with 9 seeds in each row. How many plants will be in the garden?

Ⓐ 32 seeds
Ⓑ 36 seeds
Ⓒ 42 seeds
Ⓓ 13 seeds

7. Mrs. Huerta's class is having a pizza party. There are 24 students in the class. Each pizza has 12 slices. How many pizzas does Mrs. Huerta need to order for each child to have 1 slice?

Ⓐ 3 pizzas
Ⓑ 2 pizzas
Ⓒ 1 pizza
Ⓓ 4 pizzas

8. There are 27 apples. How many pies can be made if each pie uses 3 apples?

Ⓐ 7 pies
Ⓑ 8 pies
Ⓒ 9 pies
Ⓓ 10 pies

9. Keegan is planting a garden in even rows. He has 48 seeds. Which layout is NOT possible?

Ⓐ 6 rows of 8 seeds
Ⓑ 8 rows of 6 seeds
Ⓒ 7 rows of 7 seeds
Ⓓ 12 rows of 4 seeds

10. There are 25 students in a gym class. They want to play a game with 5 equal teams. How many students will be on each team?

Ⓐ 4 students
Ⓑ 5 students
Ⓒ 7 students
Ⓓ 3 students

© Lumos Information Services 2018 | LumosLearning.com

Name ______________________________ Date ________________

11. Josie has 7 days to read a book with 21 chapters. How many chapters should she read each day?

Ⓐ 3 chapters
Ⓑ 4 chapters
Ⓒ 5 chapters
Ⓓ 7 chapters

12. Devon has $40 to spend on fuel. One gallon of fuel costs $5. How many gallons can Devon afford to buy?

Ⓐ 5 gallons
Ⓑ 12 gallons
Ⓒ 9 gallons
Ⓓ 8 gallons

13. Amanda is using the following cake recipe:
4 cups flour
1 cup sugar
3 cups milk
1 egg
If Amanda needs to make three batches, how many cups of flour will she need?

Ⓐ 7 cups
Ⓑ 12 cups
Ⓒ 10 cups
Ⓓ 16 cups

14. Kim invited 20 friends to her birthday party. Twice as many friends than she invited showed up the day of the party. Which number sentence could be used to solve how many friends came to the party?

Ⓐ $n + 20 = 2$
Ⓑ $n \times 20 = 2$
Ⓒ $20 \times 2 = n$
Ⓓ $20 - n = 20$

15. The product of 9 and a number is 45.
Which number sentence models this situation?

Ⓐ $9 + n = 45$
Ⓑ $45 + 9 = n$
Ⓒ $9 \times n = 45$
Ⓓ $5 \times n = 45$

© Lumos Information Services 2018 | LumosLearning.com

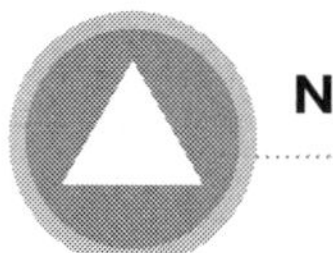

Name ______________________ Date ____________

Chapter 1

Lesson 4: Finding Unknown Values

You can scan the QR code given below or use the url to access additional EdSearch resources including videos and mobile apps related to *Finding Unknown Values*.

edSearch **_Finding Unknown Values_**

URL	QR Code
http://www.lumoslearning.com/a/3oaa4	

© Lumos Information Services 2018 | LumosLearning.com

Name ______________________________ Date ________________

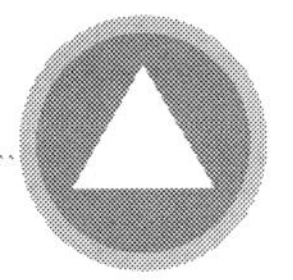

1. Find the number that makes this equation true.
   n x 6 = 30

   Ⓐ n = 11
   Ⓑ n = 7
   Ⓒ n = 5
   Ⓓ n = 3

2. Find the number that makes this equation true.
   7 x ___ = 21

   Ⓐ 3
   Ⓑ 4
   Ⓒ 5
   Ⓓ 6

3. Find the number that makes this equation true.
   ___ x 4 = 36

   Ⓐ 9
   Ⓑ 8
   Ⓒ 7
   Ⓓ 6

4. Find the number that makes this equation true.
   n ÷ 9 = 8

   Ⓐ n = 81
   Ⓑ n = 45
   Ⓒ n = 72
   Ⓓ n = 63

5. Find the number that makes this equation true.
   ____ ÷ 3 = 10

   Ⓐ 27
   Ⓑ 30
   Ⓒ 33
   Ⓓ 60

© Lumos Information Services 2018 | LumosLearning.com

6. Find the number that makes this equation true.
   $45 \div n = 9$

   Ⓐ n = 10
   Ⓑ n = 7
   Ⓒ n = 5
   Ⓓ n = 3

7. Find the number that makes this equation true.
   64 = ___ x 8

   Ⓐ 6
   Ⓑ 7
   Ⓒ 8
   Ⓓ 9

8. Find the number that makes this equation true.
   12 ÷ ___ = 2

   Ⓐ 7
   Ⓑ 6
   Ⓒ 8
   Ⓓ 4

9. Find the number that makes this equation true.
   ___ ÷ 7 = 11

   Ⓐ 63
   Ⓑ 70
   Ⓒ 77
   Ⓓ 78

10. Find the number that makes this equation true.
    $16 = n \times 4$

    Ⓐ n = 12
    Ⓑ n = 4
    Ⓒ n = 3
    Ⓓ n = 2

© Lumos Information Services 2018 | LumosLearning.com

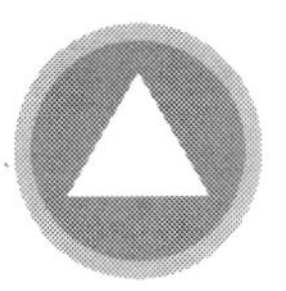

11. For what value of m is this equation true?
    $m \times 7 = 56$

    Ⓐ m = 6
    Ⓑ m = 8
    Ⓒ m = 7
    Ⓓ m = 12

12. For what value of n is this equation true?
    $60 \div n = 5$

    Ⓐ n = 8
    Ⓑ n = 12
    Ⓒ n = 14
    Ⓓ n = 16

13. For what value of z is this equation true?
    $9 \times z = 81$

    Ⓐ z = 11
    Ⓑ z = 9
    Ⓒ z = 12
    Ⓓ z = 7

14. For what value of p is this equation true?
    $p \div 3 = 3$

    Ⓐ p = 6
    Ⓑ p = 1
    Ⓒ p = 9
    Ⓓ p = 0

15. For what value of u is this equation true?
    $10 \div u = 10$

    Ⓐ u = 1
    Ⓑ u = 0
    Ⓒ u = 10
    Ⓓ u = 100

© Lumos Information Services 2018 | LumosLearning.com

Name ______________________ Date ______________

Chapter 1

Lesson 5: Multiplication & Division Properties

You can scan the QR code given below or use the url to access additional EdSearch resources including videos and mobile apps related to *Multiplication & Division Properties*.

edSearch ***Multiplication & Division Properties***

URL	QR Code
http://www.lumoslearning.com/a/3oab5	

© Lumos Information Services 2018 | LumosLearning.com

Name ______________________ Date ______________

1. Which of these statements is not true?

Ⓐ 4 x (3 x 6) = (4 x 3) x 6
Ⓑ 4 x 3 = 3 x 4
Ⓒ 15 x 0 = 0 x 15
Ⓓ 12 x 1 = 12 x 12

2. Which of these statements is true?

Ⓐ The product of 11 x 6 is equal to the product of 6 x 11.
Ⓑ The product of 11 x 6 is greater than the product of 6 x 11.
Ⓒ The product of 11 x 6 is less than the product of 6 x 11.
Ⓓ There is no relationship between the product of 11 x 6 and the product of 6 x 11.

3. Which of the following expressions has a value of 0?

Ⓐ (3 x 4) x 1
Ⓑ 50 x 1
Ⓒ 3 x 4 x 0
Ⓓ (3 x 1) x 2

4. Which two numerical expressions both have a value of 0?

Ⓐ 60 x 1 and 1 x 60
Ⓑ 10 x 10 and 0 x 10
Ⓒ 27 x 0 and 0 x 27
Ⓓ 0 ÷ 15 and 15 ÷ 15

5. Which mathematical property does this equation model?
6 x 1 = 6

Ⓐ Commutative Property of Multiplication
Ⓑ Associative Property of Multiplication
Ⓒ Identity Property of Multiplication
Ⓓ Distributive Property

© Lumos Information Services 2018 | LumosLearning.com

6. Which mathematical property does this equation model?
9 x 6 = 6 x 9

Ⓐ Commutative Property of Multiplication
Ⓑ Associative Property of Multiplication
Ⓒ Identity Property of Multiplication
Ⓓ Distributive Property

7. Which mathematical property does this equation model?
(2 x 10) x 3 = 2 x (10 x 3)

Ⓐ Commutative Property of Multiplication
Ⓑ Associative Property of Multiplication
Ⓒ Identity Property of Multiplication
Ⓓ Distributive Property

8. Which mathematical property does this equation model?
4 x (9 + 6) = (4 x 9) + (4 x 6)

Ⓐ Commutative Property of Multiplication
Ⓑ Associative Property of Multiplication
Ⓒ Identity Property of Multiplication
Ⓓ Distributive Property

9. By the Commutative Property of Multiplication, if you know that 4 x 5= 20, then you also know that ______ .

Ⓐ 20 is an even number
Ⓑ 4 x 6 = 24
Ⓒ 5 x 4 = 20
Ⓓ 5 is greater than 4

10. By the Associative Property of Multiplication, If you know that (2 x 3) x 4 = 24, then you also know that _________.

Ⓐ 2 x (3 x 4) = 24
Ⓑ 2 x 4 = 8
Ⓒ 24 ÷ 6 = 4
Ⓓ (2 x 3) x 5 = 30

© Lumos Information Services 2018 | LumosLearning.com

Name ______________________________ Date _______________

11. Complete the following statement:
Multiplication and _______ are inverse operations.

Ⓐ addition
Ⓑ subtraction
Ⓒ division
Ⓓ distribution

12. 32 x 7 = 7 x 32
This equation models the __________.

Ⓐ Commutative Property of Multiplication
Ⓑ Associative Property of Multiplication
Ⓒ Identity Property of Multiplication
Ⓓ Distributive Property

13. 26 x 2 = (20 x 2) + (6 x 2)
This equation models the __________ .

Ⓐ Commutative Property of Multiplication
Ⓑ Associative Property of Multiplication
Ⓒ Identity Property of Multiplication
Ⓓ Distributive Property

14. By the Identity Property of Multiplication, you know that _________.

Ⓐ 2 x 2 = 4
Ⓑ 0 x 0 = 0
Ⓒ 6 x 1 = 6
Ⓓ 5 ÷ 5 = 1

15. What number belongs in the blank?
10 x __ = 10

Ⓐ 1
Ⓑ 0
Ⓒ 10
Ⓓ 5

© Lumos Information Services 2018 | LumosLearning.com

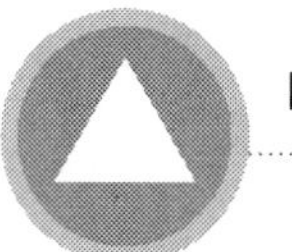

Name ______________________ Date ______________

Chapter 1

Lesson 6: Relating Multiplication & Division

You can scan the QR code given below or use the url to access additional EdSearch resources including videos and mobile apps related to Relating *Multiplication & Division*.

edSearch	Relating Multiplication & Division
URL	QR Code
http://www.lumoslearning.com/a/3oab6	

© Lumos Information Services 2018 | LumosLearning.com

1. Find the number that would complete both of the following number sentences.
___ x 6 = 30
30 ÷ 6 = ___

Ⓐ 7
Ⓑ 5
Ⓒ 6
Ⓓ 24

2. Find the number that would complete both of the following number sentences.
7 x ___ = 21
21 ÷ ___ = 7

Ⓐ 5
Ⓑ 14
Ⓒ 3
Ⓓ 7

3. Find the number that would complete both of the following number sentences.
72 ÷ ___ = 8
8 x ___ = 72

Ⓐ 8
Ⓑ 9
Ⓒ 10
Ⓓ 64

4. Find the number that would complete both of the following number sentences.
50 ÷ ___ = 5
___ x 5 = 50

Ⓐ 15
Ⓑ 5
Ⓒ 45
Ⓓ 10

© Lumos Information Services 2018 | LumosLearning.com

5. Find the number that would complete both of the following number sentences.
36 ÷ ____ =
_____ x 9 = 36

Ⓐ 4
Ⓑ 5
Ⓒ 27
Ⓓ 9

6. There are 9 students in a group. Each student needs 5 sheets of paper to complete a project. Which number sentence below can be used to find out how many total sheets of paper are needed for this project?

Ⓐ 45 x ____ = 9
Ⓑ 9 x 45 = ____
Ⓒ ____ ÷ 5 = 9
Ⓓ 9 ÷ 5 = ____

7. In a football game, Timmy scored 8 touchdowns. Each touchdown was worth 7 points. Which number sentence below can be used to find out how many points Timmy scored in all?

Ⓐ 56 x ___ = 8
Ⓑ ____ ÷ 7 = 8
Ⓒ 7 x 56 = ____
Ⓓ 8 ÷ 7 = _____

8. Devon needs to buy 96 pencils for his goodie bags. Pencils are sold in packages of 12. Which number sentence below can be used to find out how many packages Devon needs to buy?

Ⓐ 12 ÷ 96 = ___
Ⓑ 12 x 96 = ____
Ⓒ ____ x 12 = 96
Ⓓ 12 ÷ ____ = 96

© Lumos Information Services 2018 | LumosLearning.com

Name ________________________ Date ______________

9. Eighty-four students are attending an awards ceremony. They are to be seated at twelve equal tables. Which number sentence below can be used to find out how many students should be assigned to each table?

Ⓐ 6 x ____ = 84
Ⓑ 12 x ____ = 84
Ⓒ 84 x 12 = _____
Ⓓ 12 ÷ 84 = ____

10. Walter has 16 slices of pizza to share among himself and seven friends. He wants each person to get an equal number of slices. Which number sentence below can be used to find out how many slices each person will get?

Ⓐ 7 ÷ 16 = ____
Ⓑ 7 x 16 = ____
Ⓒ 7 x ____ = 16
Ⓓ 8 x ____ = 16

11. Which of the following equations would be in the same fact family as:
6 x 5 = 30?

Ⓐ 30 ÷ 10 = 3
Ⓑ 6 x 30 = 5
Ⓒ 30 ÷ 5 = 6
Ⓓ 5 ÷ 30 = 6

12. Which number sentence is equivalent to the number sentence below?
4 x n = 32

Ⓐ 4 x 32 = n
Ⓑ n + 4 = 32
Ⓒ 32 ÷ 4 = n
Ⓓ 4 x 4 = n

13. Which number sentence is equivalent to the number sentence below?
n x 6 = 48

Ⓐ 48 x n = 6
Ⓑ 6 x 4 = n
Ⓒ 48 x 6 = n
Ⓓ 48 ÷ n = 6

© Lumos Information Services 2018 | LumosLearning.com

14. Which number sentence is equivalent to the number sentence below?
45 ÷ n = 9

Ⓐ **9 x 45 = n**
Ⓑ **45 x n = 9**
Ⓒ **9 x n = 45**
Ⓓ **n + 9 = 45**

15. David receives 2 pieces of candy for each chore that he completes each week. This week he earned 32 pieces of candy. Which number sentence below can be used to figure out how many chores David completed?

Ⓐ **2 x 32 = ___**
Ⓑ **___ x 2 = 32**
Ⓒ **32 + 2 = ___**
Ⓓ **2 ÷ 32 = ___**

© Lumos Information Services 2018 | LumosLearning.com

Name ______________________ Date ______________

Chapter 1

Lesson 7: Multiplication & Division Facts

You can scan the QR code given below or use the url to access additional EdSearch resources including videos and mobile apps related to *Multiplication & Division Facts*.

edSearch **Multiplication & Division Facts**

URL	QR Code
http://www.lumoslearning.com/a/3oac7	

© Lumos Information Services 2018 | LumosLearning.com

Name ______________________________ Date ________________

1. Find the product.
   6 x 0 = ____

   Ⓐ 6
   Ⓑ 1
   Ⓒ 0
   Ⓓ 2

2. Find the product.
   1 x 10 = ____

   Ⓐ 0
   Ⓑ 1
   Ⓒ 10
   Ⓓ 11

3. Solve.
   3 x 8 = ____

   Ⓐ 24
   Ⓑ 21
   Ⓒ 18
   Ⓓ 28

4. Solve.
   ___ = 5 x 9

   Ⓐ 40
   Ⓑ 45
   Ⓒ 50
   Ⓓ 35

5. Find the product of 8 and 6.

   Ⓐ 14
   Ⓑ 42
   Ⓒ 48
   Ⓓ 56

© Lumos Information Services 2018 | LumosLearning.com

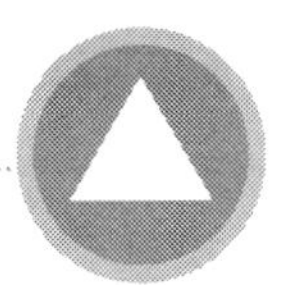

6. Find the product of 7 and 7.

Ⓐ 42
Ⓑ 46
Ⓒ 49
Ⓓ 56

7. Find the product of 4 and 6.

Ⓐ 20
Ⓑ 24
Ⓒ 28
Ⓓ 32

8. Find the product.
6 x 9 = ____

Ⓐ 54
Ⓑ 45
Ⓒ 48
Ⓓ 64

9. Find the product.
___ = 9 x 8

Ⓐ 64
Ⓑ 72
Ⓒ 81
Ⓓ 82

10. Which expression below has a product of 48?

Ⓐ 6 x 7
Ⓑ 4 x 14
Ⓒ 7 x 8
Ⓓ 8 x 6

11. Find the quotient of 25 and 5.

Ⓐ 20
Ⓑ 5
Ⓒ 4
Ⓓ 15

© Lumos Information Services 2018 | LumosLearning.com

12. What is 32 divided by 4?

Ⓐ 9
Ⓑ 8
Ⓒ 7
Ⓓ 6

13. What is 28 divided by 7?

Ⓐ 4
Ⓑ 5
Ⓒ 3
Ⓓ 6

14. Find the quotient.
$0 \div 5 =$ ____

Ⓐ 0
Ⓑ 1
Ⓒ 5
Ⓓ 50

15. Find the quotient.
$7 \div 1 =$ ____

Ⓐ 1
Ⓑ 0
Ⓒ 7
Ⓓ 8

16. Find the quotient.
_____ $= 12 \div 2$

Ⓐ 9
Ⓑ 8
Ⓒ 7
Ⓓ 6

17. Divide.
$63 \div 9 =$ ____

Ⓐ 6
Ⓑ 7
Ⓒ 8
Ⓓ 9

© Lumos Information Services 2018 | LumosLearning.com

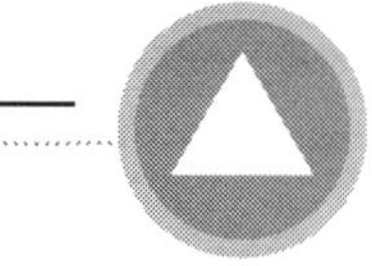

18. Divide.
42 ÷ 7 = ____

Ⓐ 5
Ⓑ 6
Ⓒ 7
Ⓓ 8

19. Find the quotient of 33 and 3.

Ⓐ 11
Ⓑ 12
Ⓒ 10
Ⓓ 9

20. Divide.
56 ÷ 7 = ____

Ⓐ 6
Ⓑ 7
Ⓒ 8
Ⓓ 9

21. Solve.
4 x 12 = _____

Ⓐ 36
Ⓑ 48
Ⓒ 42
Ⓓ 46

22. Solve.
____ = 75 ÷ 5

Ⓐ 13
Ⓑ 15
Ⓒ 17
Ⓓ 25

23. Solve.
84 ÷ 12 = _____

Ⓐ 7
Ⓑ 8
Ⓒ 9
Ⓓ 12

© Lumos Information Services 2018 | LumosLearning.com

24. Solve.
12 x 3 = ____

Ⓐ **32**
Ⓑ **36**
Ⓒ **39**
Ⓓ **48**

25. Solve.
36 ÷ 3 = _____

Ⓐ **22**
Ⓑ **12**
Ⓒ **14**
Ⓓ **18**

26. Solve.
60 ÷ 5 = ____

Ⓐ **8**
Ⓑ **9**
Ⓒ **12**
Ⓓ **14**

27. Solve.
11 x 4 = ____

Ⓐ **32**
Ⓑ **44**
Ⓒ **39**
Ⓓ **46**

28. Solve.
_____ = 80 ÷ 8

Ⓐ **10**
Ⓑ **9**
Ⓒ **8**
Ⓓ **12**

© Lumos Information Services 2018 | LumosLearning.com

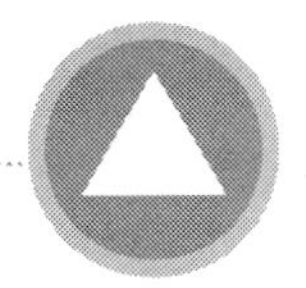

29. Solve.
12 x 8 = ____

Ⓐ 72
Ⓑ 84
Ⓒ 92
Ⓓ 96

30. Solve.
50 ÷ 5 = ____

Ⓐ 25
Ⓑ 20
Ⓒ 10
Ⓓ 50

© Lumos Information Services 2018 | LumosLearning.com

Name ______________________ Date ______________

Chapter 1

Lesson 8: Two-Step Problems

You can scan the QR code given below or use the url to access additional EdSearch resources including videos and mobile apps related to *Two-Step Problems*.

edSearch ***Two-Step Problems***

URL	QR Code
http://www.lumoslearning.com/a/3oad8	

© Lumos Information Services 2018 | LumosLearning.com

Name ______________________________ Date ________________

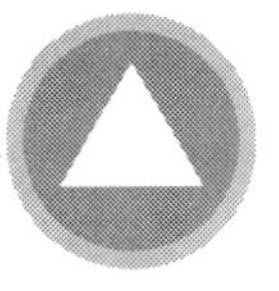

1. Danny has 47 baseball cards. He gives his brother 11 cards. Danny then divides the remaining cards between 3 of his classmates. How many cards does each classmate receive?

Ⓐ 15
Ⓑ 3
Ⓒ 12
Ⓓ 11

2. Two third grade classes are lined up outside. One class is lined up in 3 rows of 7. The other class is lined up in 4 rows of 5. How many total third graders are lined up outside?

Ⓐ 19 third graders
Ⓑ 21 third graders
Ⓒ 41 third graders
Ⓓ 20 third graders

3. Jessica earns 10 dollars per hour for babysitting. She has saved 60 dollars so far. How many more hours will she need to babysit to buy something that costs 100 dollars?

Ⓐ 40 hours
Ⓑ 6 hours
Ⓒ 10 hours
Ⓓ 4 hours

4. George started with 2 bags of 10 cookies. He gave 12 cookies to his parents. How many cookies does George have now?

Ⓐ 8 cookies
Ⓑ 10 cookies
Ⓒ 12 cookies
Ⓓ 20 cookies

5. Renae has 60 minutes to do her chores and do her homework. She has 3 chores to complete and each chore takes 15 minutes to complete. After completing her chores, how many minutes does Renae have left to do her homework?

Ⓐ 15 minutes
Ⓑ 45 minutes
Ⓒ 30 minutes
Ⓓ 0 minutes

© Lumos Information Services 2018 | LumosLearning.com

6. Anna and Jamie want to buy a new board game. The original cost was 28 dollars. It is on sale for 4 dollars off. How much money should each girl pay if they buy the game on sale and pay equal amounts?

Ⓐ $24
Ⓑ $2
Ⓒ $12
Ⓓ $14

7. 100 students went on a field trip. Ten students rode with their parents in a car while the remaining students were divided equally on 5 buses. How many students rode on each bus?

Ⓐ 9 students
Ⓑ 18 students
Ⓒ 50 students
Ⓓ 90 students

8. Julia has 32 books. Her sister has twice the number of books that Julia has. How many books do the girls have altogether?

Ⓐ 66 books
Ⓑ 32 books
Ⓒ 64 books
Ⓓ 96 books

9. Alicia bought 5 crates of apples. Each crate had 8 apples. She divided the apples equally into 10 bags. How many apples were in each bag?

Ⓐ 40 apples
Ⓑ 4 apples
Ⓒ 10 apples
Ⓓ 5 apples

10. Janeth went to the store and spent 4 dollars on markers. She also bought 3 copies of the same book. If she spent a total of 19 dollars, how much did each book cost?

Ⓐ 5 dollars
Ⓑ 4 dollars
Ⓒ 3 dollars
Ⓓ 6 dollars

© Lumos Information Services 2018 | LumosLearning.com

Name ______________________________ Date ______________

11. Brian won 24 candy bars in a contest. He gave 2 candy bars to each of his 7 friends. How many candy bars does Brian have left?

Ⓐ 14 candy bars
Ⓑ 12 candy bars
Ⓒ 10 candy bars
Ⓓ 17 candy bars

12. Jenine gave 3 mini cupcakes to each of her three sisters. She then had 4 left for herself. How many mini cupcakes did Jenine start?

Ⓐ 13 mini cupcakes
Ⓑ 9 mini cupcakes
Ⓒ 10 mini cupcakes
Ⓓ 7 mini cupcakes

13. Twenty-two people visited the art exhibit at the museum on Friday. Twice as many people visited on Saturday. How many people combined visited the art exhibit at the museum on Friday and Saturday

Ⓐ 88 people
Ⓑ 66 people
Ⓒ 22 people
Ⓓ 44 people

14. Audrey can watch 5 hours of TV a week. She has already watched 4 shows that are each 1 hour long. How many more hours can she watch TV this week?

Ⓐ 3 hours
Ⓑ 2 hours
Ⓒ 1 hour
Ⓓ 4 hours

15. Greg had 3 books. His older brother gave him 15 more books. Greg wants to divide his total number of books equally onto 6 shelves. How many books should he place on each shelf?

Ⓐ 3 books
Ⓑ 12 books
Ⓒ 18 books
Ⓓ 6 books

© Lumos Information Services 2018 | LumosLearning.com

Name ______________________ Date ______________

Chapter 1

Lesson 9: Number Patterns

You can scan the QR code given below or use the url to access additional EdSearch resources including videos and mobile apps related to *Number Patterns*.

edSearch ***Number Patterns***

URL	QR Code
http://www.lumoslearning.com/a/3oad9	

Name ________________________________ Date ________________

1. Which of the following is an even number?

Ⓐ 764,723
Ⓑ 90,835
Ⓒ 5,862
Ⓓ 609

2. Which of these sets contains no odd numbers?

Ⓐ 13, 15, 81, 109, 199
Ⓑ 123, 133, 421, 412, 600
Ⓒ 34, 46, 48, 106, 88
Ⓓ 12, 37, 6, 14, 144

3. Complete the following statement.
The sum of two even numbers will always be ______ .

Ⓐ greater than 10
Ⓑ less than 100
Ⓒ even
Ⓓ odd

4. Complete the following statement.
The product of two even numbers will always be _____ .

Ⓐ even
Ⓑ odd
Ⓒ a multiple of 10
Ⓓ a square number

5. Complete the following statement.
A number has a nine in its ones place. The number must be a multiple of ____.

Ⓐ 9
Ⓑ 3
Ⓒ 7
Ⓓ None of the above

© Lumos Information Services 2018 | LumosLearning.com

6. Complete the following statement.
   Numbers that are multiples of 8 are all _______.

   Ⓐ even
   Ⓑ multiples of 2
   Ⓒ multiples of 4
   Ⓓ All of the above

7. If this pattern continues, what will the next 3 numbers be?
   7, 14, 21, 28, 35, ...

   Ⓐ 41, 47, 53
   Ⓑ 49, 56, 63
   Ⓒ 77, 84, 91
   Ⓓ 42, 49, 56

8. Complete the following statement.
   All multiples of _____ can be decomposed into two equal addends.

   Ⓐ 6
   Ⓑ 9
   Ⓒ 3
   Ⓓ 5

9. If this pattern continues, what will the next 3 numbers be?
   9, 18, 27, 36, ...

   Ⓐ 54, 63, 72
   Ⓑ 45, 54, 63
   Ⓒ 44, 52, 60
   Ⓓ 44, 53, 62

10. A number is multiplied by 7 and the resulting product is even. Which of these could have been the number?

   Ⓐ 7
   Ⓑ 17
   Ⓒ 34
   Ⓓ 99

© Lumos Information Services 2018 | LumosLearning.com

Name ______________________ Date ______________

11. Complete the following statement.
The multiples of 4 will always ______.

Ⓐ have a 2 in the ones place
Ⓑ be even
Ⓒ be divisible by 8
Ⓓ None of these

12. Complete the following statement.
The sum of an even number and an odd number will always be _______.

Ⓐ even
Ⓑ odd
Ⓒ divisible by 3
Ⓓ None of the above

13. Complete the following statement.
A multiple of 4 can have a _____ in its ones place.

Ⓐ 2
Ⓑ 8
Ⓒ 6
Ⓓ All of the above

14. Complete the following statement.
A multiple of 5 can have a _____ as its ones digit.

Ⓐ 0
Ⓑ 3
Ⓒ 9
Ⓓ All of the above

15. Which of the following would produce an even product?

Ⓐ an even number times an even number
Ⓑ an even number times an odd number
Ⓒ an odd number times an even number
Ⓓ All of the above

End of Operations and Algebraic Thinking

© Lumos Information Services 2018 | LumosLearning.com

Chapter 1:

Operations and Algebraic Thinking

Answer Key
&
Detailed Explanations

© Lumos Information Services 2018 | LumosLearning.com

Name ______________________ Date ______________

Lesson 1: Understanding Multiplication

Question No.	Answer	Detailed Explanation
1	D	The picture depicts 3 sets of 9 objects which is equivalent to 3 x 9 = 27.
2	C	The picture depicts 4 sets of 5 objects which is equivalent to 4 x 5.
3	D	The picture depicts 4 sets of 7 objects which is equivalent to 4 x 7 = 28.
4	A	The picture depicts 2 sets of 12 objects which is equivalent to 2 x 12 = 24.
5	A	The picture depicts 4 sets of 4 objects which is equivalent to 4 x 4 = 16.
6	A	The picture depicts 4 sets of 6 objects which is equivalent to 4 x 6 = 24.
7	D	The picture depicts 3 sets of 6 objects which is equivalent to 6 x 3 (or 3 x 6) = 18
8	D	The picture depicts 5 sets of 3 objects which is equivalent to 3 x 5 (or 5 x 3) = 15.
9	A	The picture depicts 2 sets of 3 objects which is equivalent to 3 x 2 (or 2 x 3) = 6.
10	C	The picture depicts 4 sets of 5 objects which is equivalent to 5 x 4 (or 4 x 5) = 20.
11	B	The picture depicts 4 sets of 2 objects which is equivalent to 4 x 2 = 8.
12	B	The picture depicts 6 sets of 8 objects which is equivalent to 6 x 8 = 48.
13	C	The picture depicts 2 sets of 10 objects which is equivalent to 2 x 10 = 20.
14	A	The picture depicts 5 sets of 5 objects which is equivalent to 5 x 5 = 25.
15	D	The picture depicts 3 sets of 3 objects which is equivalent to 3 x 3 = 9.

© Lumos Information Services 2018 | LumosLearning.com

Name ______________________ Date ____________

Lesson 2: Understanding Division

Question No.	Answer	Detailed Explanation
1	D	There are 100 items that need to be divided into 10 groups. $100 \div 10 = 10$.
2	C	There are 15 items that need to be divided into 3 groups. $15 \div 3 = 5$.
3	A	There are 50 items that need to be divided into 5 groups. $50 \div 5 = 10$.
4	D	There are 30 items that need to be sorted into 6 groups. $30 \div 6 = 5$.
5	B	There are 80 items that need to be sorted into groups of 10. $80 \div 10 = 8$.
6	B	There are 15 items that need to be shared with 5 groups. $15 \div 5 = 3$.
7	D	There are 81 items that need to be divided into 9 groups. $81 \div 9 = 9$.
8	C	There are 40 items that need to be sorted into groups of 10. $40 \div 10 = 4$.
9	B	There are 20 items that need to be sorted into groups of 5. $20 \div 5 = 4$.
10	A	There are 30 people who need to be divided into 5 groups. $30 \div 5 = 6$.
11	D	There are 24 items that need to be sorted into groups of 3. $24 \div 3 = 8$.
12	B	There are 44 items that need to be divided into 4 groups. $44 \div 4 = 11$.
13	C	All together, the group has $32 + 4$ marbles which equals 36. There are 4 people in the group. There are 36 items that need to be shared with 4 groups. $36 \div 4 = 9$.
14	B	There are 100 items (dollars) that need to be divided into 5 groups. $\$100 \div 5 = \$20$.
15	D	There are 26 items that need to be shared equally between two people. $26 \div 2 = 13$.

© Lumos Information Services 2018 | LumosLearning.com

Name ______________________ Date ______________

Lesson 3: Applying Multiplication & Division

Question No.	Answer	Detailed Explanation
1	B	Product refers to the result of the multiplication of two or more numbers. 54 and 3 are both factors.
2	D	The phrase "twice as many" indicates that if a number is multiplied by 2, the product will reflect two times, or twice, the original amount. In this case, the product of 30 popcorn balls is already known. The product must be then divided by 2 in order to find the amount of cotton candy. $30 \div 2 = 15$.
3	C	The phrase "8 times as many" indicates that if Sue's amount of DVDs is multiplied by 8, the product will be equal to the amount of Monica's DVD's. To solve for Sue use the equation $n \times 8 = 56$. When trying to solve for a missing number in a multiplication equation, you must divide the product by the given number. $56 \div 8 = 7$
4	A	Marcus' jumping jack is equivalent to 4 times that of Jonathan. $7 \times 4 = 28$.
5	D	Marsha's movie count is equivalent to 3 times that of Jonathan's. $3 \times 4 = 12$.
6	B	There are 4 groups and each group has 9 items. This indicates that if the number of groups is multiplied by the number of items in each group, the product will reflect the total number of items in all. $4 \times 9 = 36$.
7	B	The number of students can be divided by the number of slices in each pizza. The quotient will reflect the amount of pizzas needed. $24 \div 12 = 2$ pizzas.
8	C	The total number of apples can be divided by the amount of apples needed for each pie. The quotient will reflect the amount of pies that can be made. $27 \div 3 = 9$.
9	C	The number of rows multiplied by the number of seeds in each row must equal 48. All answer choices given contain two numbers with a product of 48 except Option C. $7 \times 7 = 49$.
10	B	The total number of students can be divided by the number of teams. The quotient will reflect the amount of students on each team. $25 \div 5 = 5$.

© Lumos Information Services 2018 | LumosLearning.com

Question No.	Answer	Detailed Explanation
11	A	The number of chapters can be divided by the number of days. The quotient will reflect the amount of chapters that can be read each day. 21 ÷ 7 = 3.
12	D	The $40 Devon has can be divided by the cost of one gallon of fuel ($5). The quotient will reflect the number of gallons Devon can afford to by. $40 ÷ $5 = 8 gallons.
13	B	There are 4 cups needed for one batch. Amanda is making 3 batches, so she needs 4 sets of 3 cups or 4 x 3 = 12 cups.
14	C	The phrase "twice as many" indicates that if a number is multiplied by 2 the product will reflect two times, or twice, the original amount. The amount of friends invited (20) must be multiplied by 2 in order to find out how many friends attended. The correct number sentence is 20 x 2 = 40.
15	C	Product refers to multiplication. The problem states that 9 and a number (n) when multiplied together equals 45. The correct number sentence is 9 x n = 45.

Lesson 4: Finding Unknown Values

Question No.	Answer	Detailed Explanation
1	C	To solve for an unknown in a multiplication problem, you must do the opposite operation, which is to divide. You must divide the product by the given factor. 30 ÷ 6=5. n = 5.
2	A	To solve for an unknown in a multiplication problem, you must do the opposite operation, which is to divide. You must divide the product by the given factor. 21 ÷ 7=3. The missing value is 3.
3	A	To solve for an unknown in a multiplication problem, you must do the opposite operation, which is to divide. You must divide the product by the given factor. 36 ÷ 4= 9. The missing value is 9.
4	C	The first step to solve for an unknown in a division problem is to decide which part of the problem is missing: Dividend: n Divisor: 9 Quotient: 8 When solving for the dividend, you must multiply the divisor and the quotient. 9 x 8 = 72. n = 72.
5	B	The first step to solve for an unknown in a division problem is to decide which part of the problem is missing: Dividend: n Divisor: 3 Quotient: 10 When solving for the dividend, you must multiply the divisor and the quotient. 3 x 10 = 30. n = 30.

© Lumos Information Services 2018 | LumosLearning.com

Question No.	Answer	Detailed Explanation
6	C	The first step to solve for an unknown in a division problem is to decide which part of the problem is missing: Dividend: 45 Divisor: n Quotient: 9 When solving for the divisor, you must divide the dividend by the quotient. 45 ÷ 9 = 5.
7	C	To solve for an unknown in a multiplication problem, you must do the opposite operation, which is to divide. You must divide the product by the given factor. 64 ÷ 8 = 8. n = 8.
8	B	The first step to solve for an unknown in a division problem is to decide which part of the problem is missing: Dividend: 12 Divisor: (?) Quotient: 2 When solving for the divisor, you must divide the dividend by the quotient. 12 ÷ 2 = 6. The unknown value is 6.
9	C	The first step to solve for an unknown in a division problem is to decide which part of the problem is missing: Dividend: (?) Divisor: 7 Quotient: 11 When solving for the dividend, you must multiply the divisor and the quotient. 7 x 11 = 77. The missing value is 77.
10	B	To solve for an unknown in a multiplication problem, you must do the opposite operation, which is to divide. You must divide the product by the given factor. 16 ÷ 4 = 4. n = 4.
11	B	To solve for an unknown in a multiplication problem, you must do the opposite operation, which is to divide. You must divide the product by the given factor. 56 ÷ 7= 8. m = 8.
12	B	The first step to solve for an unknown in a division problem is to decide which part of the problem is missing: Dividend: 60 Divisor: n Quotient: 5 When solving for the divisor, you must divide the dividend by the quotient. 60 ÷ 5 = 12. The unknown value is 12.

© Lumos Information Services 2018 | LumosLearning.com

Question No.	Answer	Detailed Explanation
13	B	To solve for an unknown in a multiplication problem, you must do the opposite operation, which is to divide. You must divide the product by the given factor. 81 ÷ 9= 9. z = 9.
14	C	The first step to solve for an unknown in a division problem is to decide which part of the problem is missing: Dividend: n Divisor: 3 Quotient: 3 When solving for the dividend, you must multiply the divisor and the quotient. 3 x 3 = 9. p = 9.
15	A	The first step to solve for an unknown in a division problem is to decide which part of the problem is missing: Dividend: 10 Divisor: n Quotient: 10 When solving for the divisor, you must divide the dividend by the quotient. 10 ÷ 10 = 1. n = 1.

Lesson 5: Multiplication & Division Properties

Question No.	Answer	Detailed Explanation
1	D	In order for the statement to be true, the answer on both sides of the equal sign must be the same. All of the answer choices are equal except for the last choice. 12 x 1 = 12 and 12 x 12 = 144. 12 x 1 is not equal to 12 x 12.
2	A	Option A is the only one that is true because 11 x 6 = 66 and 6 x 11 = 66. This is an example of the Commutative Property of Multiplication.
3	C	In multiplication, the only time that the number 0 will be the product is when at least one of the factors is 0. Option C is the only choice that fits this rule.
4	C	In multiplication, the only time that the number 0 will be the product is when at least one of the factors is 0. In division, the only time that the number 0 will be the answer is when the dividend is 0. Option C is the only choice where the equation fits this rule.

© Lumos Information Services 2018 | LumosLearning.com

Name ______________________ Date ______________

Question No.	Answer	Detailed Explanation
5	C	The Identity Property of Multiplication states that any number multiplied by 1 equals itself.
6	A	The Commutative Property of Multiplication states that the order of the factors does not change the answer.
7	B	The Associative Property of Multiplication states that the grouping of factors does not change the answer.
8	D	The Distributive Property states that multiplying a number by a group of numbers added together is the same as doing each multiplication problem separately.
9	C	The Commutative Property of Multiplication states that the order of the factors does not change the answer. 4 x 5 = 20 is the same as 5 x 4 = 20.
10	A	The Associative Property of Multiplication states that the grouping of factors does not change the answer. So (2 x 3) x 4 = 24 is the same as 2 x (3 x 4) = 24.
11	C	Inverse operations means that one operation will reverse the effect of another. Division is the inverse of multiplication and vice versa. For example, if 4 x 3 = 12 then 12 ÷ 3 = 4 and 12 ÷ 4 = 3.
12	A	The Commutative Property of Multiplication states that the order of the factors does not change the answer.
13	D	The Distributive Property states that multiplying a number by a group of numbers added together is the same as doing each multiplication separately.
14	C	The Identity Property of Multiplication states that any number multiplied by 1 equals itself.
15	A	The Identity Property of Multiplication states that any number multiplied by 1 equals itself. 10 x 1 = 10.

© Lumos Information Services 2018 | LumosLearning.com

Name ______________________ Date ______________

Lesson 6: Relating Multiplication & Division

Question No.	Answer	Detailed Explanation
1	B	5 x 6 = 30 is equivalent to 30 ÷ 6 = 5 because 5 groups of 6 objects is equivalent to 30.
2	C	7 x 3 = 21 is equivalent to 21 ÷ 3 = 7 because 7 groups of 3 objects is equivalent to 21.
3	B	8 x 9 = 72 is equivalent to 72 ÷ 9 = 8 because 8 groups of 9 objects is equivalent to 72.
4	D	10 x 5 = 50 is equivalent to 50 ÷ 10 = 5 because 10 groups of 5 objects is equivalent to 50.
5	A	4 x 9 = 36 is equivalent to 36 ÷ 4 = 9 because 4 groups of 9 objects is equivalent to 36.
6	C	Since there are 9 students who need 5 sheets each, this is equivalent to 9 x 5 which equals 45, or 45 ÷ 5 = 9.
7	B	Since there are 8 touchdowns each worth 7 points, this is equivalent to 8 x 7 which equals 56, or 56 ÷ 7 = 8.
8	C	There are 12 items in a pack and there are 96 items needed. To see how many packs are needed, you must divide 96 ÷ 12 = 8, or 8 x 12 = 96.
9	B	There are 12 tables and 84 students which means there are 84 students who need to be divided into groups of 12, which is 84 ÷ 12 = 7 or 12 x 7 = 84.
10	D	There are 8 people, Walter plus his 7 friends who need to evenly split 16 slices of pizza which is equivalent to 16 ÷ 8 = 2, or 8 x 2 = 16.
11	C	Equations in a fact family utilize the same numbers in a multiplication or division problem and are equivalent. 6 x 5 = 30 is equivalent to 30 ÷ 5 = 6.
12	C	Multiplication and division are inverse operations. This means that when two numbers are multiplied, the product can be divided by either of the two factors to give the other factor. 4 x n =32 is equivalent to 32 ÷ 4 = n
13	D	Multiplication and division are inverse operations. This means that when two numbers are multiplied, the product can be divided by either of the two factors to give the other factor. n x 6 = 48 is equivalent to 48 ÷ n = 6

© Lumos Information Services 2018 | LumosLearning.com

Question No.	Answer	Detailed Explanation
14	C	Multiplication and division are inverse operations. This means that when two numbers are multiplied, the product can be divided by either of the two factors to give the other factor. 45 ÷ n = 9 is equivalent to 9 x n = 45.
15	B	There are 32 items that need to be evenly split into 2 groups which is equivalent to 32 ÷ 2 = 16, or 16 x 2 =32.

Lesson 7: Multiplication & Division Facts

Question No.	Answer	Detailed Explanation
1	C	In multiplication, any time one of the factors is 0, the product is also 0.
2	C	The Identity Property of Multiplication states that any number multiplied by 1 equals itself, number so 1 x 10 = 10.
3	A	3 x 8 represents 3 groups of 8 items. There are 24 items total.
4	B	5 x 9 represents 5 groups of 9 items. There are 45 items total
5	C	Product refers to the answer when numbers are multiplied. 8 x 6 represents 8 groups of 6 items. There are 48 items total.
6	C	Product refers to the answer when numbers are multiplied. 7 x 7 represents 7 groups of 7 items. There are 49 items total.
7	B	Product refers to the answer when numbers are multiplied. 4 x 6 represents 4 groups of 6 items. There are 24 items total.
8	A	Product refers to the answer when numbers are multiplied. 6 x 9 represents 6 groups of 9 items. There are 54 items total.
9	B	Product refers to the answer when numbers are multiplied. 9 x 8 represents 9 groups of 8 items. There are 72 items total.
10	D	Product refers to the answer when numbers are multiplied. Option D is the only choice in which the answer is 48.
11	B	The quotient refers to the answer when a number is divided by another number. There are 25 items that need to be divided into 5 groups. 25 ÷ 5 = 5.

© Lumos Information Services 2018 | LumosLearning.com

Question No.	Answer	Detailed Explanation
12	B	There are 32 items that need to be divided into 4 groups. 32 ÷ 4 = 8
13	A	There are 28 items that need to be divided into 7 groups. 28 ÷ 7 = 4. The quotient refers to the answer when a number is divided by another number.
14	A	When the number 0 is divided by any non-zero number, the answer is always 0. The quotient refers to the answer when a number is divided by another number.
15	C	When a number is divided by 1, the answer is always the original number.
16	D	The quotient refers to the answer when a number is divided by another number. There are 12 items that need to be divided into 2 groups. 12 ÷ 2 = 6.
17	B	There are 63 items that need to be divided into 9 groups. 63 ÷ 9 = 7.
18	B	There are 42 items that need to be divided into 7 groups. 42 ÷ 7 = 6
19	A	The quotient refers to the answer when a number is divided by another number. There are 33 items that need to be divided into 3 groups. 33 ÷ 3 = 11.
20	C	There are 56 items that need to be divided into 7 groups. 56 ÷ 7 = 8.
21	B	4 x 12 represents 4 groups of 12 items. There are 48 items total.
22	B	There are 75 items that need to be divided into 5 groups. 75 ÷ 5 = 15
23	A	There are 84 items that need to be divided into 12 groups. 84 ÷ 12 = 7.
24	B	12 x 3 represents 12 groups of 3 items. There are 36 items total.
25	B	There are 36 items that need to be divided into 3 groups. 36 ÷ 3 = 12.
26	C	There are 60 items that need to be divided into 5 groups. 60 ÷ 5 = 12.
27	B	11 x 4 represents 11 groups of 4 items. There are 44 items total.

© Lumos Information Services 2018 | LumosLearning.com

Question No.	Answer	Detailed Explanation
28	A	There are 80 items that need to be divided into 8 groups. 80 ÷ 8 = 10.
29	D	12 x 8 represents 12 groups of 8 items. There are 96 items total.
30	C	There are 50 items that need to be divided into 5 groups. 50 ÷ 5 = 10.

Lesson 8: Two-Step Problems

Question No.	Answer	Detailed Explanation
1	C	First, calculate how many cards Danny has by subtracting 11 from 47; 47 - 11 = 36. Then divide this number by 3 to see how many each classmate will receive; 36 ÷ 3 = 12.
2	C	First, calculate how many students are in each class. The first class has 3 rows of 7 students and 3 x 7 = 21. The second class has 4 rows of 5 students and 4 x 5 = 20. Then add both totals to calculate the total number of students outside; 21 + 20 = 41.
3	D	First, calculate how much more money Jessica needs to save by subtracting what she has from what she needs; $100 - $60 = $40. Jessica needs 40 more dollars. Now divide 40 dollars by the amount she makes each hour of babysitting to find how many more hours she needs to work to earn the rest of the money; $40 ÷ 10 = 4.
4	A	First calculate how many cookies George began with by multiplying 2 and 10; 2 x 10 = 20. Then subtract the number of cookies he gave to his parents from this total; 20 - 12 = 8.
5	A	First calculate the total number of minutes Renae spends doing chores by multiplying 3 and 15; 3 x 15 = 45. Then subtract this number from 60 minutes to see how many minutes she has remaining to do her homework; 60 - 45 = 15.
6	C	First calculate the sale price of the game; $28 - 4 = $24. Then divide this answer by 2 to see how much each girl will pay; $24 ÷ 2 = $12.

© Lumos Information Services 2018 | LumosLearning.com

Name ____________________ Date ____________

Question No.	Answer	Detailed Explanation
7	B	First, calculate how many students rode the bus by subtracting the number of students who rode in cars from the total number of students; 100 - 10 = 90. Then divide this answer by the number of buses to see how many students rode on each bus; 90 ÷ 5 = 18.
8	D	First, calculate the number of books the sister has by multiplying the number of books Julia has by 2; 32 x 2 = 64. Then add this number to Julia's amount to get the total books; 64 + 32 = 96.
9	B	First, calculate the total number of apples Alicia has by multiplying 5 and 8; 5 x 8 = 40. Then calculate the number of apples in each bag by dividing the total number of apples by the number of bags; 40 ÷ 10 = 4.
10	A	First, subtract the cost of the markers from the total amount spent; $19 - 4 = $15. Then divide this answer by the number of books bought to calculate the cost of each book; $15 ÷ 3 = $5.
11	C	First, calculate how many candy bars were given to friends by multiplying the total number of friends by the number of bars each friend received; 7 x 2 = 14. Then subtract this answer from the total number of candy bars Brian won; 24 - 14 = 10.
12	A	First, calculate how many mini cupcakes were given to the sisters by multiplying the total number of sisters by the number each one received; 3 x 3 = 9. Then add this number to the number of cupcakes Jenine had left for herself; 9 + 4 = 13.
13	B	First, calculate how many people visited the museum on Saturday by multiplying the number of Friday visitors by 2 (for twice as many); 22 x 2 = 44. Then add this number to the number of Friday visitors; 44 + 22 = 66.
14	C	First, calculate how many hours Audrey has already watched TV by multiplying the number of shows she has watched by the length of each show; 4 x 1 = 4. Then subtract this answer from the total number of hours she is allowed to watch; 5 - 4 = 1.
15	A	First, calculate the total number of books Greg has by adding the number of books given to him by his brother to the number of books he already had; 15 + 3 = 18. Then calculate the number of books on each shelf by dividing the total number of books by the number of shelves; 18 ÷ 6 = 3.

© Lumos Information Services 2018 | LumosLearning.com

Name ______________________ Date ______________

Lesson 9: Number Patterns

Question No.	Answer	Detailed Explanation
1	C	An even number is any number whose ones digit is one of the following numbers: 0, 2, 4, 6, 8. Option C is the only choice that fits this criteria.
2	C	An odd number is any number whose ones digit is one of the following numbers: 1, 3, 5, 7, 9. Option C is the only choice that does not contain any numbers that fit this criteria.
3	C	The rule states that when two even numbers are added, the answer will always be even. For example, 34 + 12 = 46.
4	A	The rule states that when two even numbers are multiplied, the product will always be even. For example, 34 x 4 = 136.
5	D	There is not enough information given for us to decide if the number is a multiple of 3, 7, or 9. For example, if the original number was 19, it would not be a multiple of 3, 7, or 9.
6	D	Any multiple of an even number is also even. Numbers that are multiples of 8 are also multiples of 2 and 4 because 2 and 4 are factors of 8.
7	D	The pattern is adding 7 to each number to make the next number (7 + 7 = 14, 14 + 7 = 21 and so on). The next number will be 35 + 7 = 42, the next will be 42 + 7 = 49, and the next will be 49 + 7 = 56. These numbers also represent the multiples of 7 in order.
8	A	"Two equal addends" means the number can be divided into two equal numbers. This can be performed on all even numbers. The number 6 is an even number and all of its multiples are also even.
9	B	The pattern is adding 9 to each number to make the next number (9 + 9 = 18, 18 + 9 = 27 and so on). The next number will be 36 + 9 = 45, the next will be 45 + 9 = 54, and the next will be 54 + 9 = 63.
10	C	If an odd number is multiplied by an even number, the answer will be an even number.
11	B	Multiples of even numbers are always even. Four is an even number, so all of its multiples are also even.

© Lumos Information Services 2018 | LumosLearning.com

Question No.	Answer	Detailed Explanation
12	B	The rule states that when an odd number is added to an even number, the answer will always be odd. For example, 45 + 18 = 63.
13	D	The multiples of 4 are {4, 8, 12, 16, 20, 24, 28, 32, . . .} It is apparent from this list that a multiple of 4 could have a 6, 8, or 2 as its ones digit.
14	A	All multiples of 5 have a number 0 or 5 in the ones position. For example, 205 and 210 are both multiples of 5.
15	D	Whenever an even number is multiplied by any number, the answer will always be even.

© Lumos Information Services 2018 | LumosLearning.com

Name ______________________ Date ______________

Chapter 2: Number & Operations in Base Ten

Lesson 1: Rounding Numbers

You can scan the QR code given below or use the url to access additional EdSearch resources including videos and mobile apps related to *Rounding Numbers*.

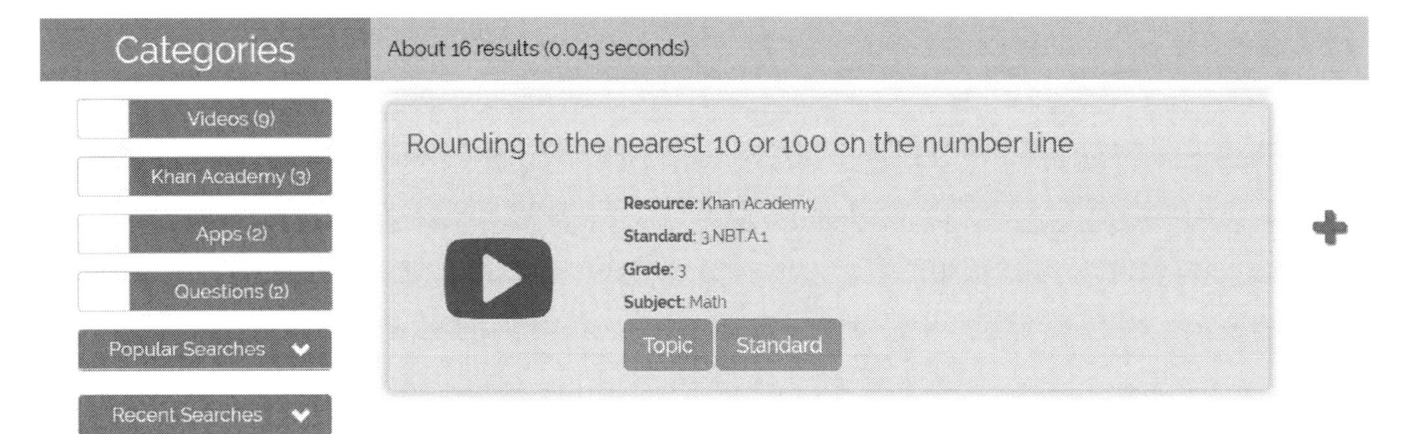

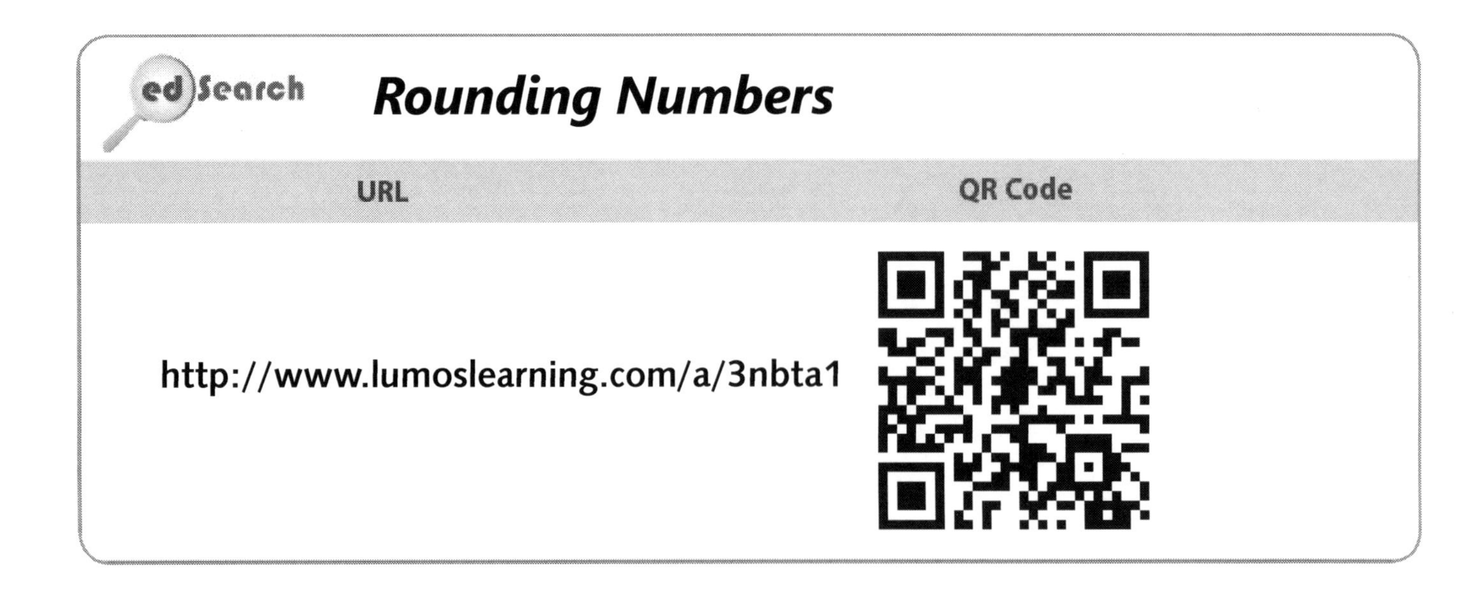

© Lumos Information Services 2018 | LumosLearning.com

Name ______________________________ Date ________________

1. **What is the value of the 9 in 11,291?**

Ⓐ 9 ones
Ⓑ 9 hundreds
Ⓒ 9 thousands
Ⓓ 9 tens

2. **What is the value of the digit 6 in 36,801?**

Ⓐ Six thousand
Ⓑ Sixty
Ⓒ Sixty thousand
Ⓓ Six hundred

3. **Which of these numbers has a 9 in the thousands place?**

Ⓐ 690,099
Ⓑ 900
Ⓒ 209,866
Ⓓ 90,786

4. **Round 2,564 to the nearest hundred.**

Ⓐ 2,000
Ⓑ 2,500
Ⓒ 2,600
Ⓓ 2,700

5. **Round 1,043 to the nearest hundred.**

Ⓐ 1,000
Ⓑ 1,100
Ⓒ 1,040
Ⓓ 1,200

6. **Round 537 to the nearest ten.**

Ⓐ 500
Ⓑ 540
Ⓒ 550
Ⓓ 530

© Lumos Information Services 2018 | LumosLearning.com

Name ______________________________ Date ________________

7. Round 957 to the nearest ten.

Ⓐ 960
Ⓑ 950
Ⓒ 900
Ⓓ 1,000

8. Maya is buying pencils for the school. Maya needs to buy enough pencils for 388 students. What is this number rounded to the nearest hundred?

Ⓐ 390
Ⓑ 380
Ⓒ 400
Ⓓ 500

9. Ninety-seven chairs are needed for an audience. What is this number rounded to the nearest ten?

Ⓐ 90
Ⓑ 100
Ⓒ 80
Ⓓ 110

10. Which of the following numbers does not round to 1,000 when rounding to the nearest hundred?

Ⓐ 955
Ⓑ 1,005
Ⓒ 1,051
Ⓓ 951

11. How many whole numbers, when rounded to the nearest ten give 100 as the result?

Ⓐ 8
Ⓑ 9
Ⓒ 10
Ⓓ 11

© Lumos Information Services 2018 | LumosLearning.com

Name ______________________ Date ______________

12. Fill in the blank.
795 rounds to 800 when rounded to the nearest _______.

Ⓐ ten
Ⓑ hundred
Ⓒ ten or hundred
Ⓓ thousand

13. Fill in the blank.
1,090 rounds to 1,100 when rounded to the nearest _______.

Ⓐ ten
Ⓑ hundred
Ⓒ ten or hundred
Ⓓ thousand

14. The attendance at a local baseball game is announced to be 4,328. What is this number rounded to the nearest ten?

Ⓐ 4,300
Ⓑ 4,330
Ⓒ 4,320
Ⓓ 4,400

15. The number of plants in a garden, when rounded to the nearest hundred, rounds to 800. Which of the following could not be the number of plants in the garden?

Ⓐ 850
Ⓑ 800
Ⓒ 750
Ⓓ 849

© Lumos Information Services 2018 | LumosLearning.com

Name ______________________ Date ______________

Chapter 2

Lesson 2: Addition & Subtraction

You can scan the QR code given below or use the url to access additional EdSearch resources including videos and mobile apps related to *Addition & Subtraction*.

© Lumos Information Services 2018 | LumosLearning.com

Name ________________________ Date ______________

1. What is the standard form of 70,000 + 6,000 + 800 + 60 + 2?

Ⓐ 706,862
Ⓑ 76,862
Ⓒ 7,682
Ⓓ 782

2. Two numbers have a difference of 29. The two numbers could be ___________.

Ⓐ 11 and 18
Ⓑ 23 and 42
Ⓒ 40 and 11
Ⓓ 50 and 39

3. Two numbers add up to 756. One number is 356. What is the other number?

Ⓐ 356
Ⓑ 300
Ⓒ 400
Ⓓ 456

4. Which of these expressions has the same difference as 94 - 50?

Ⓐ 70 - 34
Ⓑ 80 - 46
Ⓒ 60 - 16
Ⓓ 90 - 54

5. Which of these number sentences is not true?

Ⓐ 88 + 12 = 90 + 10
Ⓑ 82 + 18 = 88 + 12
Ⓒ 56 + 45 = 54 + 56
Ⓓ 46 + 15 = 56 + 5

6. Jim has 640 baseball cards and 280 basketball cards. How many sports cards does Jim have in all?

Ⓐ 820 cards
Ⓑ 360 cards
Ⓒ 8,120 cards
Ⓓ 920 cards

© Lumos Information Services 2018 | LumosLearning.com

Name ________________________________ Date ______________

7. Find the difference.
860 - 659

Ⓐ 219
Ⓑ 319
Ⓒ 201
Ⓓ 19

8. The students made 565 book covers for their math books. They used up 422 of the book covers. How many book covers are left?

Ⓐ 242 book covers
Ⓑ 163 book covers
Ⓒ 987 book covers
Ⓓ 143 book covers

9. What is the difference of 32 and 5?

Ⓐ 33
Ⓑ 27
Ⓒ 160
Ⓓ 37

10. Jenny plans to sell 50 boxes of cookies to help her scout troop raise funds. She sold 20 boxes to her neighbors. Her dad sold 15 boxes at his work office. How many more boxes does she need to sell to meet her goal?

Ⓐ 15 boxes
Ⓑ 10 boxes
Ⓒ 5 boxes
Ⓓ 25 boxes

11. Sara had 124 stickers. She gave away 62 stickers and bought 73 more stickers. How many stickers does Sara have now?

Ⓐ 135 stickers
Ⓑ 62 stickers
Ⓒ 120 stickers
Ⓓ 11 stickers

© Lumos Information Services 2018 | LumosLearning.com

12. Which of these addition expressions would require regrouping of hundreds and tens?

Ⓐ 923 + 37
Ⓑ 456 + 443
Ⓒ 235 + 234
Ⓓ 576 + 442

13. There were 605 people sitting in an auditorium at the start of a show. Thirty-five people left during the intermission. How many people remained in the auditorium after the intermission?

Ⓐ 630 people
Ⓑ 580 people
Ⓒ 570 people
Ⓓ 595 people

14. If 3 tens are subtracted from 401, what is the difference?

Ⓐ 471
Ⓑ 371
Ⓒ 398
Ⓓ 381

15. Find the sum of 37 + 93 + 200.

Ⓐ 330
Ⓑ 663
Ⓒ 320
Ⓓ 300

© Lumos Information Services 2018 | LumosLearning.com

Name ______________________ Date ______________

Lesson 3: Multiplying Multiples of 10

You can scan the QR code given below or use the url to access additional EdSearch resources including videos and mobile apps related to *Multiplying Multiples of 10*.

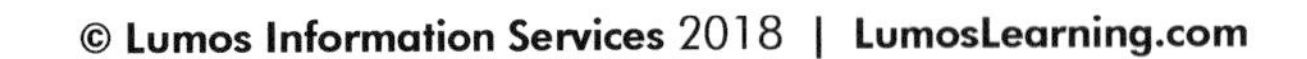
© Lumos Information Services 2018 | LumosLearning.com

Name ______________________ Date ______________

1. **Multiply:**
   6 x 10 = ____

   Ⓐ **66**
   Ⓑ **60**
   Ⓒ **61**
   Ⓓ **16**

2. **What is the product of 10 and 10?**

   Ⓐ **20**
   Ⓑ **50**
   Ⓒ **100**
   Ⓓ **1,000**

3. **Find the product.**
   5 x 40 = _____

   Ⓐ **100**
   Ⓑ **90**
   Ⓒ **200**
   Ⓓ **240**

4. **Multiply:**
   ____ = 6 x 60

   Ⓐ **120**
   Ⓑ **180**
   Ⓒ **320**
   Ⓓ **360**

5. **Find the product of 70 and 7.**

   Ⓐ **77**
   Ⓑ **140**
   Ⓒ **420**
   Ⓓ **490**

© Lumos Information Services 2018 | LumosLearning.com

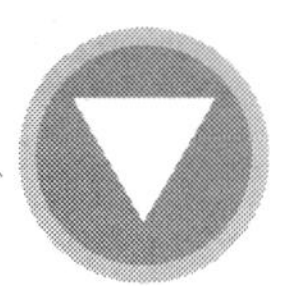

6. **Multiply:**
   _____ = 30 x 7

   Ⓐ **210**
   Ⓑ **240**
   Ⓒ **180**
   Ⓓ **100**

7. **Find the product.**
   90 x 9 = ____

   Ⓐ **800**
   Ⓑ **810**
   Ⓒ **900**
   Ⓓ **1,800**

8. **Multiply:**
   8 x 80 = _____

   Ⓐ **160**
   Ⓑ **620**
   Ⓒ **640**
   Ⓓ **660**

9. **Find the product.**
   2 x 70 = ____

   Ⓐ **140**
   Ⓑ **120**
   Ⓒ **90**
   Ⓓ **160**

10. **Multiply:**
    90 x 3 = ____

    Ⓐ **120**
    Ⓑ **180**
    Ⓒ **270**
    Ⓓ **290**

© Lumos Information Services 2018 | LumosLearning.com

Name ______________________ Date ______________

11. **Multiply:**
    2 x 10 = ____

    Ⓐ 2
    Ⓑ 5
    Ⓒ 10
    Ⓓ 20

12. **Multiply:**
    50 x 1 = ____

    Ⓐ 1
    Ⓑ 25
    Ⓒ 49
    Ⓓ 50

13. **Multiply:**
    10 x 3 = _____

    Ⓐ 10
    Ⓑ 30
    Ⓒ 100
    Ⓓ 300

14. **Find the product of 30 and 9.**

    Ⓐ 270
    Ⓑ 39
    Ⓒ 3
    Ⓓ 100

15. **Multiply:**
    10 x 7 = ____

    Ⓐ 70
    Ⓑ 10
    Ⓒ 100
    Ⓓ 7

End of Number & Operations in Base Ten

© Lumos Information Services 2018 | LumosLearning.com

Chapter 2:

Number & Operations in Base Ten

Answer Key
&
Detailed Explanations

© Lumos Information Services 2018 | LumosLearning.com

Name ______________________ Date ______________

Lesson 1: Rounding Numbers

Question No.	Answer	Detailed Explanation
1	D	Moving from right to left, the positions are as follows: ones, tens, hundreds, thousands, ten thousands. 9 - 10's is the same as $9 \times 10 = 90$.
2	A	Moving from right to left, the positions are as follows: ones, tens, hundreds, thousands, ten thousands.
3	C	Moving from right to left, the positions are as follows: ones, tens, hundreds, thousands, ten thousands.
4	C	Moving from right to left, the positions are as follows: ones, tens, hundreds, thousands. In order to round to the nearest hundred, you must look at the number in the tens place. If this number is less than 5, you must round the hundreds number down. If this number is 5 or more, you must round the hundreds number up.
5	A	Moving from right to left, the positions are as follows: ones, tens, hundreds, thousands. In order to round to the nearest hundred, you must look at the number in the tens place. If this number is less than 5, you must round the hundreds number down. If this number is 5 or more, you must round the hundreds number up.
6	B	Moving from right to left, the positions are as follows: ones, tens, hundreds. In order to round to the nearest ten, you must look at the number in the ones place. If this number is less than 5, you must round the tens number down. If this number is 5 or more, you must round the tens number up.
7	A	Moving from right to left, the positions are as follows: ones, tens, hundreds. In order to round to the nearest ten, you must look at the number in the ones place. If this number is less than 5, you must round the tens number down. If this number is 5 or more, you must round the tens number up.
8	C	Moving from right to left, the positions are as follows: ones, tens, hundreds. In order to round to the nearest hundred, you must look at the number in the tens place. If this number is less than 5, you must round the hundreds number down. If this number is 5 or more, you must round the hundreds number up.
9	B	Moving from right to left, the positions are as follows: ones, tens. In order to round to the nearest ten, you must look at the number in the ones place. If this number is less than 5, you must round the tens number down. If this number is 5 or more, you must round the tens number up.

© Lumos Information Services 2018 | LumosLearning.com

Name ______________________ Date ______________

Question No.	Answer	Detailed Explanation
10	C	Moving from right to left, the positions are as follows: ones, tens, hundreds, thousands. In order to round to the nearest hundred, you must look at the number in the tens place. If this number is less than 5, you must round the hundreds number down. If this number is 5 or more, you must round the hundreds number up. Option C is the only choice that would not round to 1,000. It would round to 1,100.
11	C	Moving from right to left, the positions are as follows: ones, tens, hundreds. In order to round to the nearest ten, you must look at the number in the ones place. If this number is less than 5, you must round the tens number down. If this number is 5 or more, you must round the tens number up. With these rules, there are 5 numbers that would round up to 100 (95, 96, 97, 98, and 99), there are 4 numbers that would round down to 100 (101,102, 103, and 104), and 100 rounds to itself. This is 10 numbers in all.
12	C	Moving from right to left, the positions are as follows: ones, tens, hundreds. In order to round to the nearest ten, you must look at the number in the ones place. If this number is less than 5, you must round the tens number down. If this number is 5 or more, you must round the tens number up. 795 has 5 in ones place. So, 795 round to 800, when rounded to nearest ten. In order to round to the nearest hundred, you must look at the number in the tens place. If this number is less than 5, you must round the hundreds number down. If this number is 5 or more, you must round the hundreds number up. 795 has a 9 in its hundreds place. So, 795 would round to 800, when rounded to nearest hundred. So, in both the cases, rounded to nearest ten or hundred, 795 would round to 800.
13	B	Moving from right to left, the positions are as follows: ones, tens, hundreds, thousands. In order to round to the nearest hundred, you must look at the number in the tens place. If this number is less than 5, you must round the hundreds number down. If this number is 5 or more, you must round the hundreds number up. 1,090 has a 9 in its tens place, so when rounding to the nearest hundred, it would round to 1,100. If 1,090 is rounded to nearest ten, it would have been the same. So, option (A) and (C) are wrong. If 1,090 is rounded to nearest thousand, it would have been 1,000. So, option (D) is also wrong.

© Lumos Information Services 2018 | LumosLearning.com

Question No.	Answer	Detailed Explanation
14	B	Moving from right to left, the positions are as follows: ones, tens, hundreds, thousands. In order to round to the nearest ten, you must look at the number in the ones place. If this number is less than 5, you must round the tens number down. If this number is 5 or more, you must round the tens number up.
15	A.	Moving from right to left, the positions are as follows: ones, tens, hundreds. In order to round to the nearest hundred, you must look at the number in the tens place. If this number is less than 5, you must round the hundreds number down. If this number is 5 or more, you must round the hundreds number up. Option A is the only choice that would not fit this criteria to round to 800.

Lesson 2: Addition & Subtraction

Question No.	Answer	Detailed Explanation
1	B	When these numbers are lined up correctly and then added, the results are: $\begin{array}{r} 70000 \\ 6000 \\ 800 \\ 60 \\ +\quad 2 \\ \hline 76,862 \end{array}$
2	C	Difference refers to the answer when two numbers are subtracted. Option C is the only choice where the difference will be 29.
3	C	In order to solve for the unknown number in an addition problem, subtract the two known numbers. 756 - 356 = 400.
4	C	The difference between 94 and 50 is 44. Option C is the only choice where 44 is also the answer.
5	C	In all of the choices, the sum on both sides of the number sentence must be equal. Option C is the only choice where the two sums are not equal. 56+45 = 101 and 54 + 56 = 110.

© Lumos Information Services 2018 | LumosLearning.com

Question No.	Answer	Detailed Explanation
6	D	The phrase "in all" indicates that the two numbers must be added in order to find the total. 640 + 280 = 920.
7	C	860 - 659 ——— 201
8	D	This problem involves subtracting the number of book covers used from the total number of book covers. 565 - 422 ——— 143
9	B	Difference refers to the answer when two numbers are subtracted. 32 - 5 ——— 27
10	A	First, add the number of cookies that were sold. 20 + 15 = 35. Then subtract this number from the amount Jenny plans to sell to calculate the number still needed to sell. 50 - 35 = 15.
11	A	First, subtract 62 from 124 to see how many stickers Sara had after giving some away. 124 - 62 = 62. Then add 62 and 73 because she bought 73 more. This will calculate the total number of stickers. 62 + 73 = 135.
12	D	Option D is the only choice that requires regrouping of the hundreds and tens place. When the tens are added, 7 + 4 = 11. One is brought down while the ten is regrouped with the hundreds.
13	C	To calculate the number of people left in the auditorium, subtract the number of people that left from the total number of people. 605 - 35 ——— 570

© Lumos Information Services 2018 | LumosLearning.com

Name ______________________ Date ______________

Question No.	Answer	Detailed Explanation
14	B	3 tens is equal to 10 + 10 + 10 = 30. 401 - 30 371
15	A	200 93 + 37 330

Lesson 3: Multiplying Multiples of 10

Question No.	Answer	Detailed Explanation
1	B	6 x 10 is equivalent to 6 sets of 10 which equals a total of 60.
2	C	Product refers to the answer when numbers are multiplied. 10 x 10 is equivalent to 10 sets of 10 which equals a total of 100.
3	C	5 x 40 is equivalent to 5 x 4 tens which equals a total of 200.
4	D	6 x 60 is equivalent to 6 x 6 tens which equals a total of 360.
5	D	Product refers to the answer when numbers are multiplied. 70 x 7 is equivalent to 7 tens x 7 which equals a total of 490.
6	A	30 x 7 is equivalent to 3 tens x 7 which equals a total of 210.
7	B	90 x 9 is equivalent to 9 tens x 9 which equals a total of 810.
8	C	8 x 80 is equivalent to 8 x 8 tens which equals a total of 640.
9	A	2 x 70 is equivalent to 2 x 7 tens which equals a total of 140.
10	C	90 x 3 is equivalent to 9 tens x 3 which equals a total of 270.
11	D	2 x 10 is equivalent to 2 sets of 10 which equals a total of 20.

© Lumos Information Services 2018 | LumosLearning.com

Question No.	Answer	Detailed Explanation
12	D	50 x 1 is equivalent to 5 tens which equals a total of 50.
13	B	10 x 3 is equivalent to 3 tens which equals a total of 30.
14	A	Product refers to the answer when numbers are multiplied. 30 x 9 is equivalent to 3 tens x 9 which equals a total of 270.
15	A	10 x 7 is equivalent to 7 tens which equals a total of 70.

© Lumos Information Services 2018 | LumosLearning.com

Name ______________________ Date ______________

Chapter 3: Number & Operations - Fractions

Lesson 1: Fractions of a Whole

You can scan the QR code given below or use the url to access additional EdSearch resources including videos and mobile apps related to *Fractions of a Whole*.

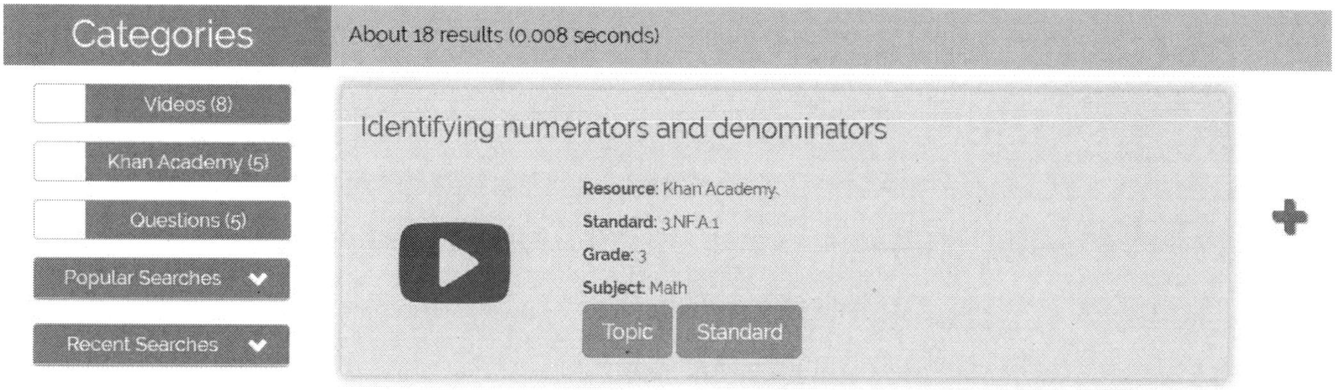

Categories

About 18 results (0.008 seconds)

Videos (8)

Khan Academy (5)

Questions (5)

Popular Searches

Recent Searches

Identifying numerators and denominators

Resource: Khan Academy

Standard: 3.NF.A.1

Grade: 3

Subject: Math

Topic | Standard

edSearch ***Fractions of a Whole***

URL	QR Code
http://www.lumoslearning.com/a/3nfa1	

© Lumos Information Services 2018 | LumosLearning.com

Name ______________________________ Date ________________

1. What fraction of the letters in the word "READING" are vowels?

Ⓐ $\frac{4}{7}$

Ⓑ $\frac{3}{4}$

Ⓒ $\frac{3}{7}$

Ⓓ $\frac{1}{3}$

2. A bag contains 3 red, 2 yellow, and 5 blue tiles. What fraction of the tiles are yellow?

Ⓐ $\frac{2}{5}$

Ⓑ $\frac{2}{10}$

Ⓒ $\frac{3}{7}$

Ⓓ $\frac{1}{3}$

3. A rectangle is cut into four equal pieces. Each piece represents what fraction of the rectangle?

Ⓐ one half
Ⓑ one third
Ⓒ one fourth
Ⓓ one fifth

4. What fraction of the square is shaded?

© Lumos Information Services 2018 | LumosLearning.com

Ⓐ $\frac{1}{2}$

Ⓑ $\frac{1}{3}$

Ⓒ $\frac{2}{1}$

Ⓓ $\frac{1}{1}$

5. What fraction of the square is shaded?

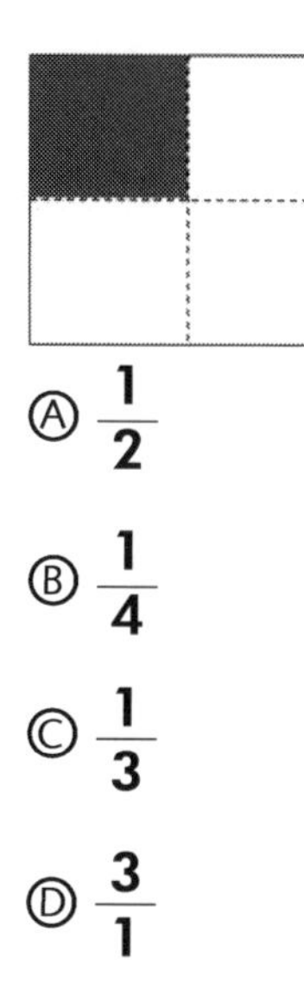

Ⓐ $\frac{1}{2}$

Ⓑ $\frac{1}{4}$

Ⓒ $\frac{1}{3}$

Ⓓ $\frac{3}{1}$

6. What fraction of the square is NOT shaded?

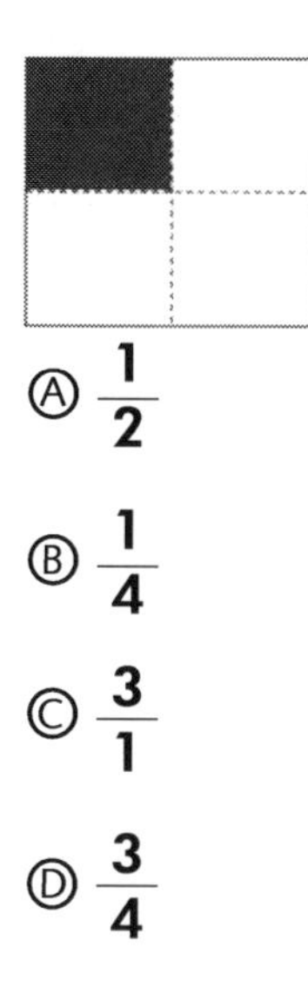

Ⓐ $\frac{1}{2}$

Ⓑ $\frac{1}{4}$

Ⓒ $\frac{3}{1}$

Ⓓ $\frac{3}{4}$

© Lumos Information Services 2018 | LumosLearning.com

7. **What fraction of the circle is shaded?**

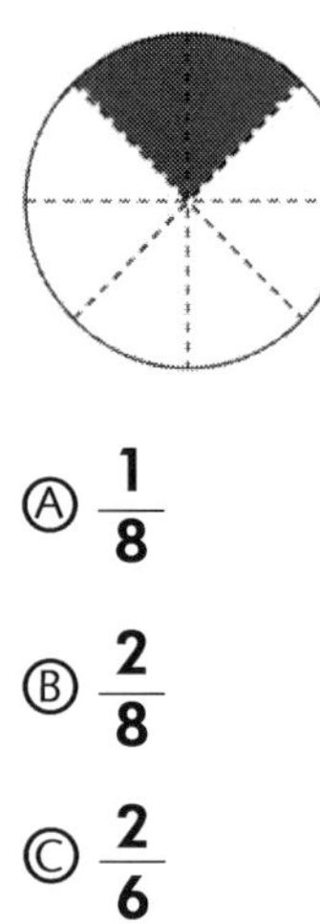

Ⓐ $\frac{1}{8}$

Ⓑ $\frac{2}{8}$

Ⓒ $\frac{2}{6}$

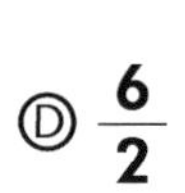

Ⓓ $\frac{6}{2}$

8. **What fraction of the circle is not shaded?**

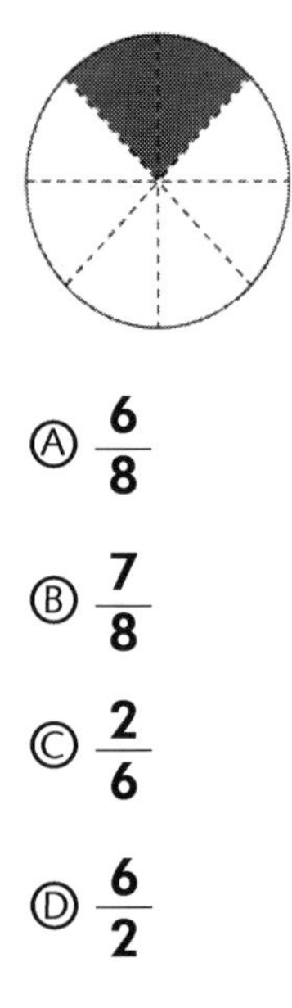

Ⓐ $\frac{6}{8}$

Ⓑ $\frac{7}{8}$

Ⓒ $\frac{2}{6}$

Ⓓ $\frac{6}{2}$

9. **What fraction of the circle is shaded?**

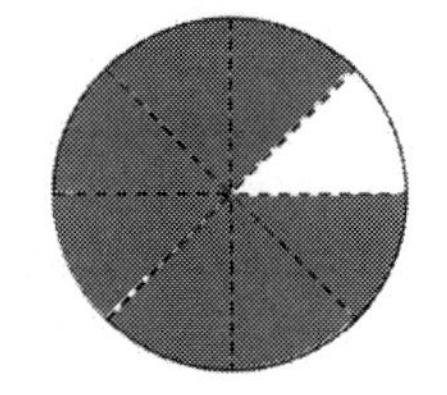

© Lumos Information Services 2018 | LumosLearning.com

Ⓐ $\frac{1}{8}$

Ⓑ $\frac{1}{7}$

Ⓒ $\frac{7}{1}$

Ⓓ $\frac{7}{8}$

10. What fraction of the circle is not shaded?

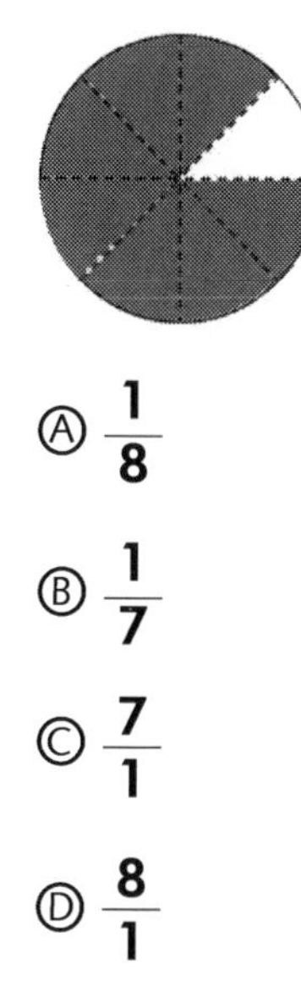

Ⓐ $\frac{1}{8}$

Ⓑ $\frac{1}{7}$

Ⓒ $\frac{7}{1}$

Ⓓ $\frac{8}{1}$

11. What fraction of the rectangle is shaded?

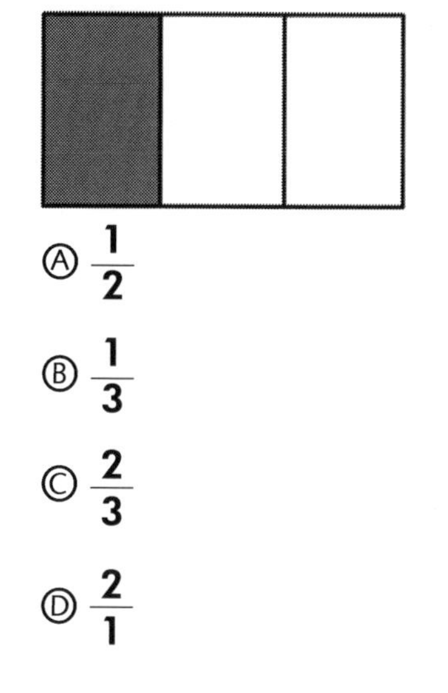

Ⓐ $\frac{1}{2}$

Ⓑ $\frac{1}{3}$

Ⓒ $\frac{2}{3}$

Ⓓ $\frac{2}{1}$

© Lumos Information Services 2018 | LumosLearning.com

12. What fraction of the rectangle is not shaded?

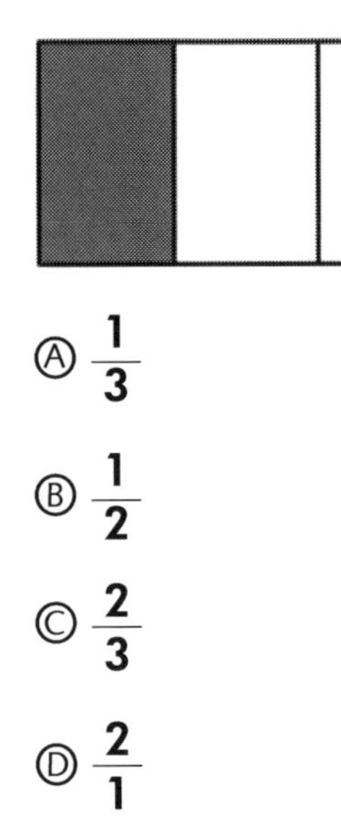

Ⓐ $\frac{1}{3}$

Ⓑ $\frac{1}{2}$

Ⓒ $\frac{2}{3}$

Ⓓ $\frac{2}{1}$

13. A pizza is cut into 12 equal slices. Eight slices are eaten. What fraction of the pizza is left?

Ⓐ $\frac{8}{12}$

Ⓑ $\frac{4}{8}$

Ⓒ $\frac{4}{12}$

Ⓓ $\frac{8}{4}$

14. The class has 20 children. Only half of the students brought their homework. How many students have their homework?

Ⓐ 20 students
Ⓑ 15 students
Ⓒ 10 students
Ⓓ 12 students

15. Meagan has 24 cupcakes. She gives a third of them to Micah. How many cupcakes does Micah have?

Ⓐ 8 cupcakes
Ⓑ 12 cupcakes
Ⓒ 3 cupcakes
Ⓓ 4 cupcakes

© Lumos Information Services 2018 | LumosLearning.com

Name ______________________ Date ______________

Chapter 3

Lesson 2: Fractions on the Number Line

You can scan the QR code given below or use the url to access additional EdSearch resources including videos and mobile apps related to *Fractions on the Number Line*.

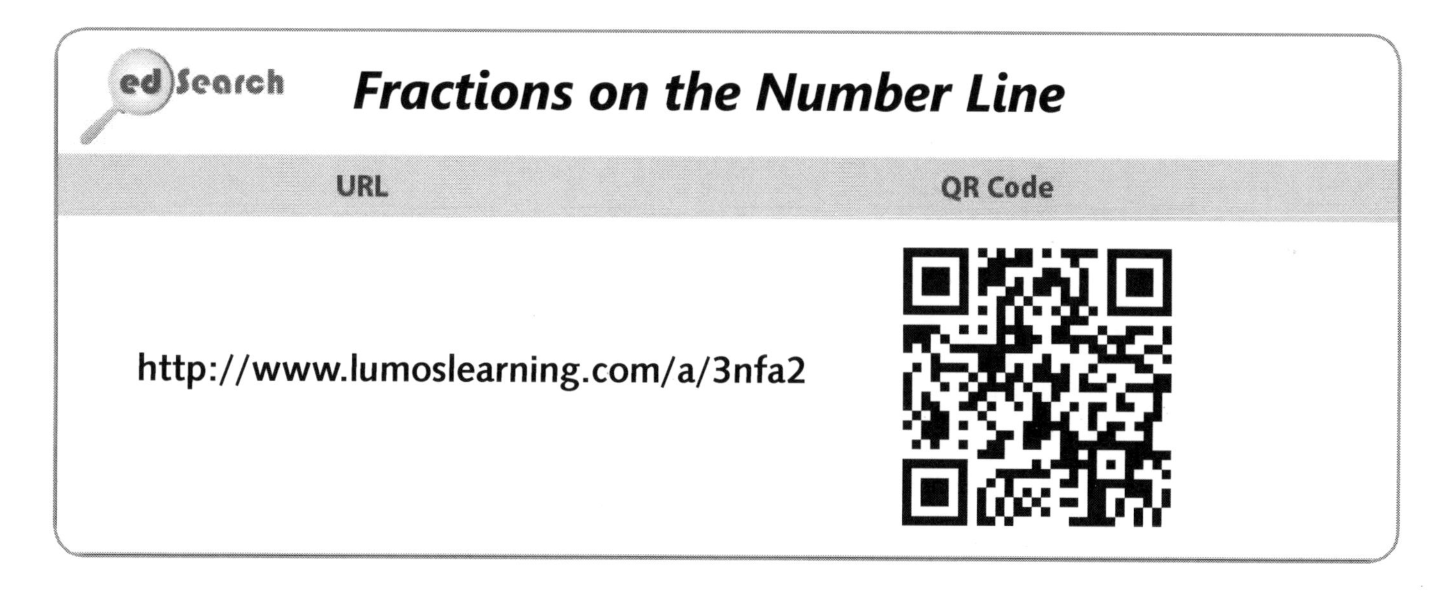

© Lumos Information Services 2018 | LumosLearning.com

1. **What fraction does the number line show?**

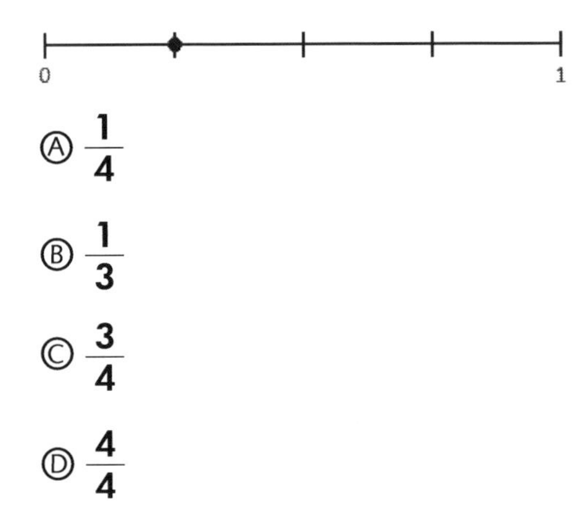

Ⓐ $\frac{1}{4}$

Ⓑ $\frac{1}{3}$

Ⓒ $\frac{3}{4}$

Ⓓ $\frac{4}{4}$

2. **What fraction does the number line show?**

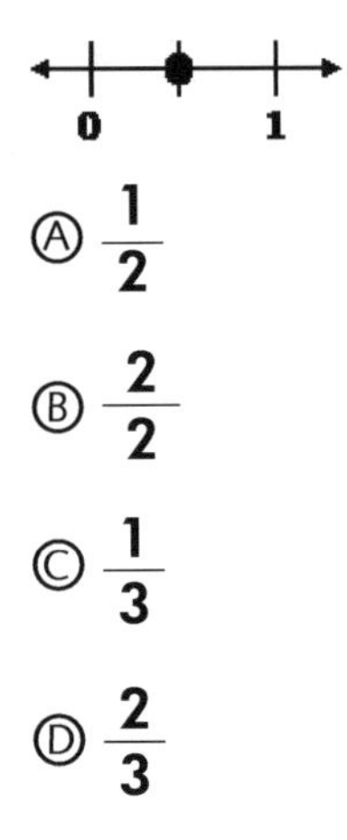

Ⓐ $\frac{1}{2}$

Ⓑ $\frac{2}{2}$

Ⓒ $\frac{1}{3}$

Ⓓ $\frac{2}{3}$

3. **What fraction does the number line show?**

0 ... 1

Ⓐ $\frac{2}{8}$

Ⓑ $\frac{3}{5}$

Ⓒ $\frac{3}{8}$

Ⓓ $\frac{4}{8}$

© Lumos Information Services 2018 | LumosLearning.com

Name ______________________ Date ______________

4. **What fraction does the number line show?**

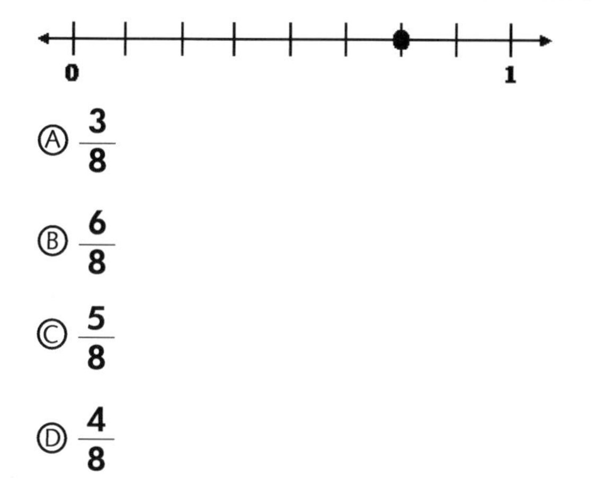

Ⓐ $\frac{3}{8}$

Ⓑ $\frac{6}{8}$

Ⓒ $\frac{5}{8}$

Ⓓ $\frac{4}{8}$

5. **What fraction does the number line show?**

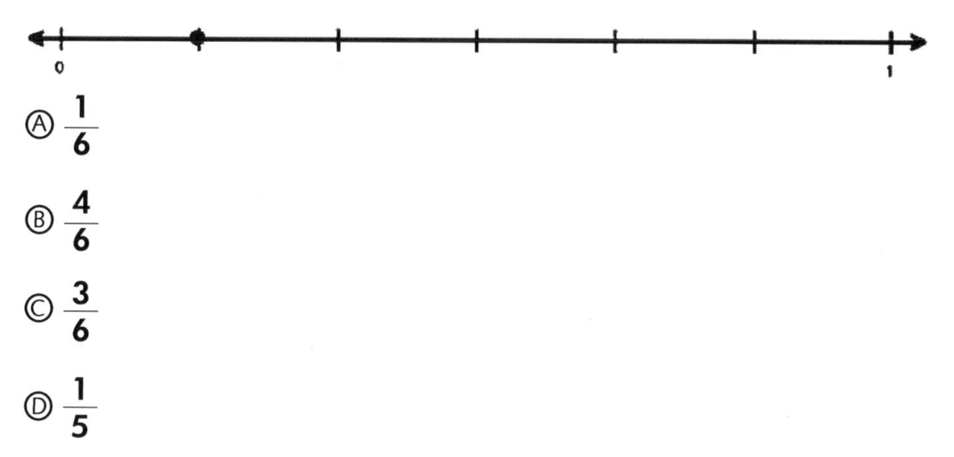

Ⓐ $\frac{1}{6}$

Ⓑ $\frac{4}{6}$

Ⓒ $\frac{3}{6}$

Ⓓ $\frac{1}{5}$

6. **What fraction does the number line show?**

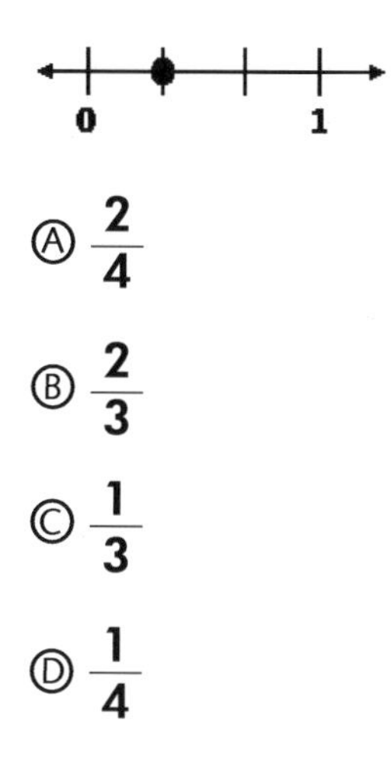

Ⓐ $\frac{2}{4}$

Ⓑ $\frac{2}{3}$

Ⓒ $\frac{1}{3}$

Ⓓ $\frac{1}{4}$

© Lumos Information Services 2018 | LumosLearning.com

7. **What fraction does the number line show?**

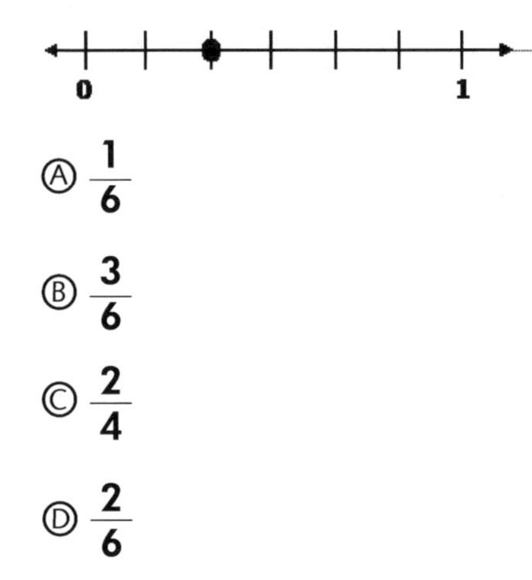

Ⓐ $\frac{1}{6}$

Ⓑ $\frac{3}{6}$

Ⓒ $\frac{2}{4}$

Ⓓ $\frac{2}{6}$

8. **What fraction does the number line show?**

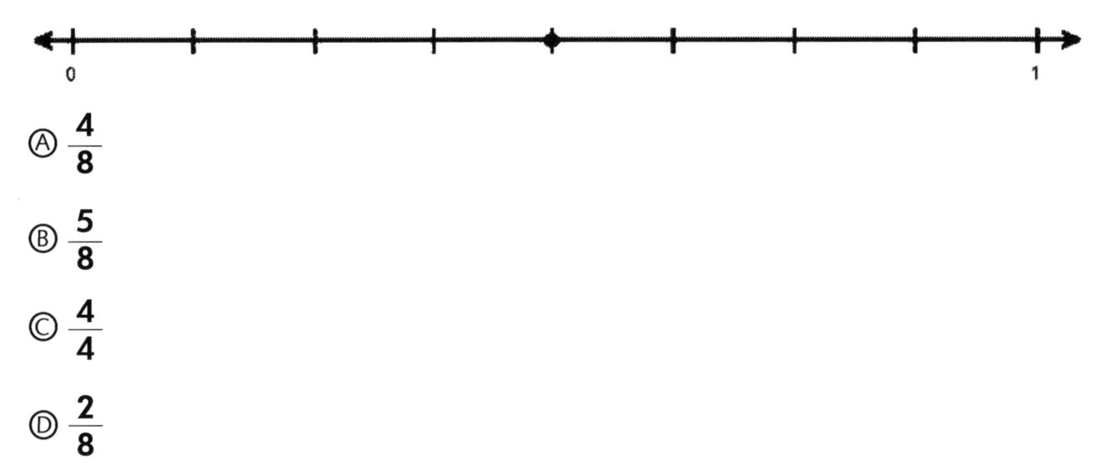

Ⓐ $\frac{4}{8}$

Ⓑ $\frac{5}{8}$

Ⓒ $\frac{4}{4}$

Ⓓ $\frac{2}{8}$

9. **What fraction does the number line show?**

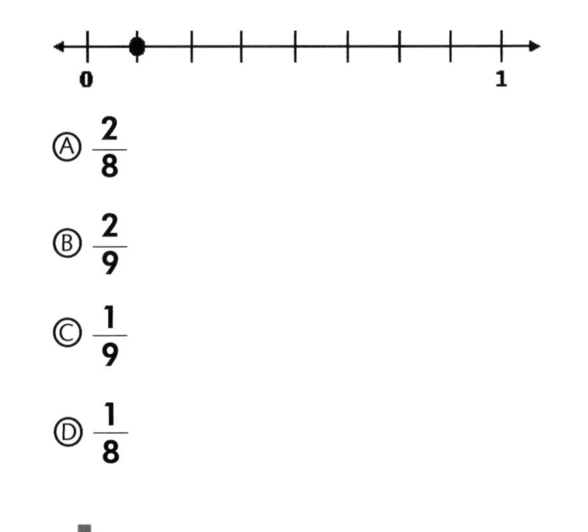

Ⓐ $\frac{2}{8}$

Ⓑ $\frac{2}{9}$

Ⓒ $\frac{1}{9}$

Ⓓ $\frac{1}{8}$

© Lumos Information Services 2018 | LumosLearning.com

Name ______________________ Date ______________

10. What fraction does the number line show

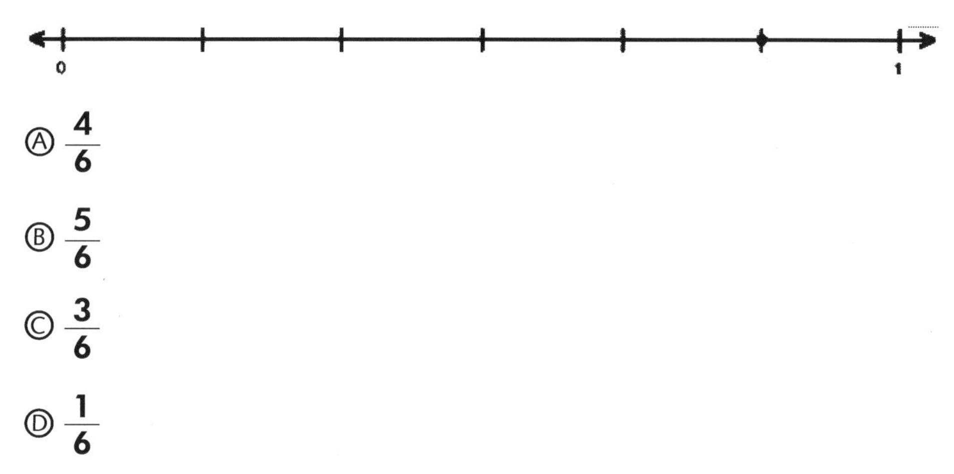

Ⓐ $\frac{4}{6}$

Ⓑ $\frac{5}{6}$

Ⓒ $\frac{3}{6}$

Ⓓ $\frac{1}{6}$

11. What fraction does the number line show?

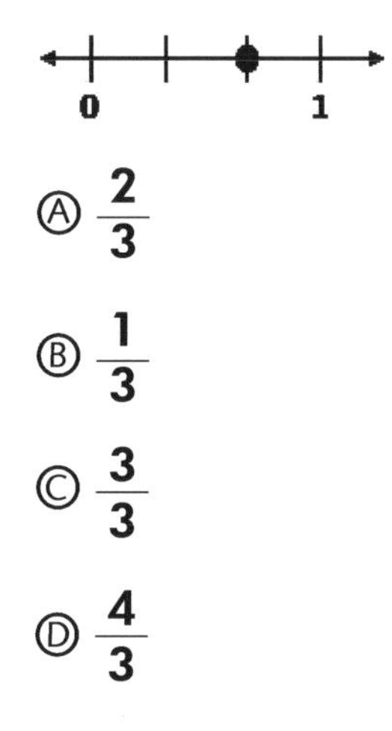

Ⓐ $\frac{2}{3}$

Ⓑ $\frac{1}{3}$

Ⓒ $\frac{3}{3}$

Ⓓ $\frac{4}{3}$

12. What fraction does the number line show?

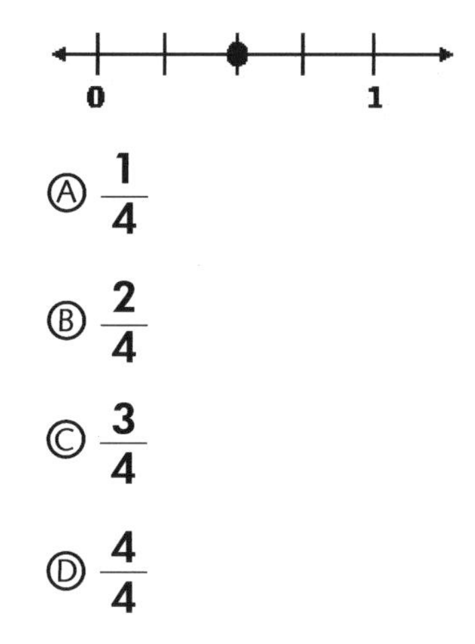

Ⓐ $\frac{1}{4}$

Ⓑ $\frac{2}{4}$

Ⓒ $\frac{3}{4}$

Ⓓ $\frac{4}{4}$

© Lumos Information Services 2018 | LumosLearning.com

13. What fraction does the number line show?

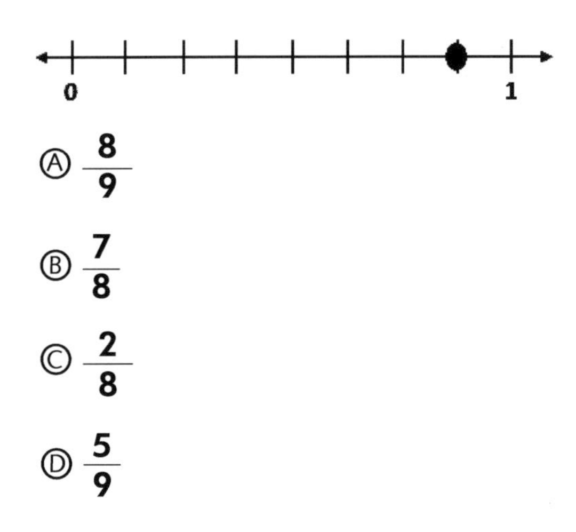

Ⓐ $\frac{8}{9}$

Ⓑ $\frac{7}{8}$

Ⓒ $\frac{2}{8}$

Ⓓ $\frac{5}{9}$

14. What fraction does the number line show?

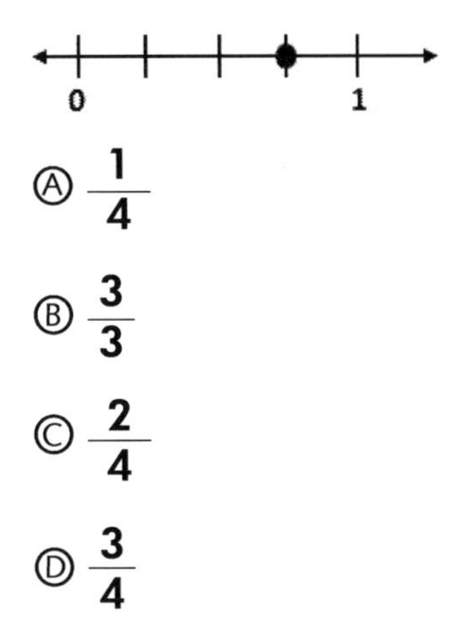

Ⓐ $\frac{1}{4}$

Ⓑ $\frac{3}{3}$

Ⓒ $\frac{2}{4}$

Ⓓ $\frac{3}{4}$

15. What fraction does the number line show?

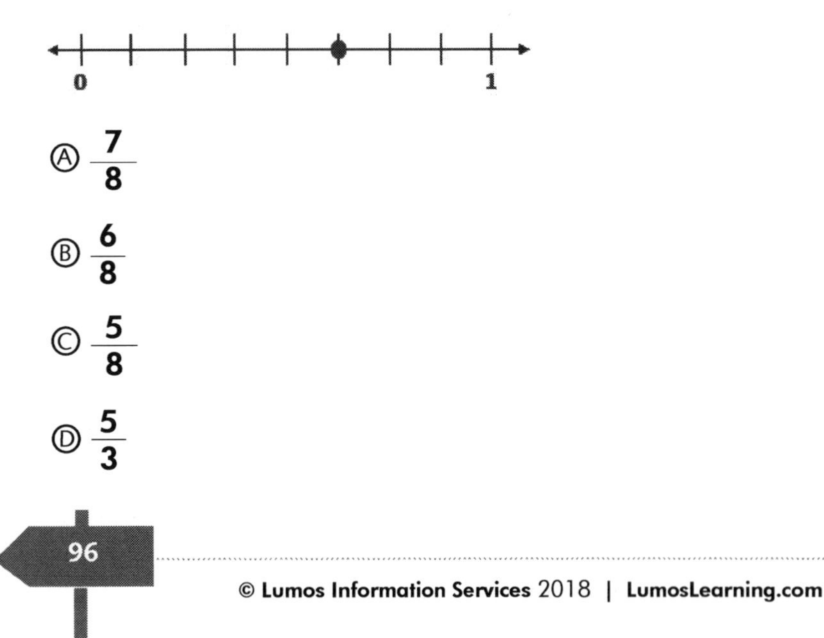

Ⓐ $\frac{7}{8}$

Ⓑ $\frac{6}{8}$

Ⓒ $\frac{5}{8}$

Ⓓ $\frac{5}{3}$

© Lumos Information Services 2018 | LumosLearning.com

Name ______________________ Date ______________

Chapter 3

Lesson 3: Comparing Fractions

You can scan the QR code given below or use the url to access additional EdSearch resources including videos and mobile apps related to *Comparing Fractions*.

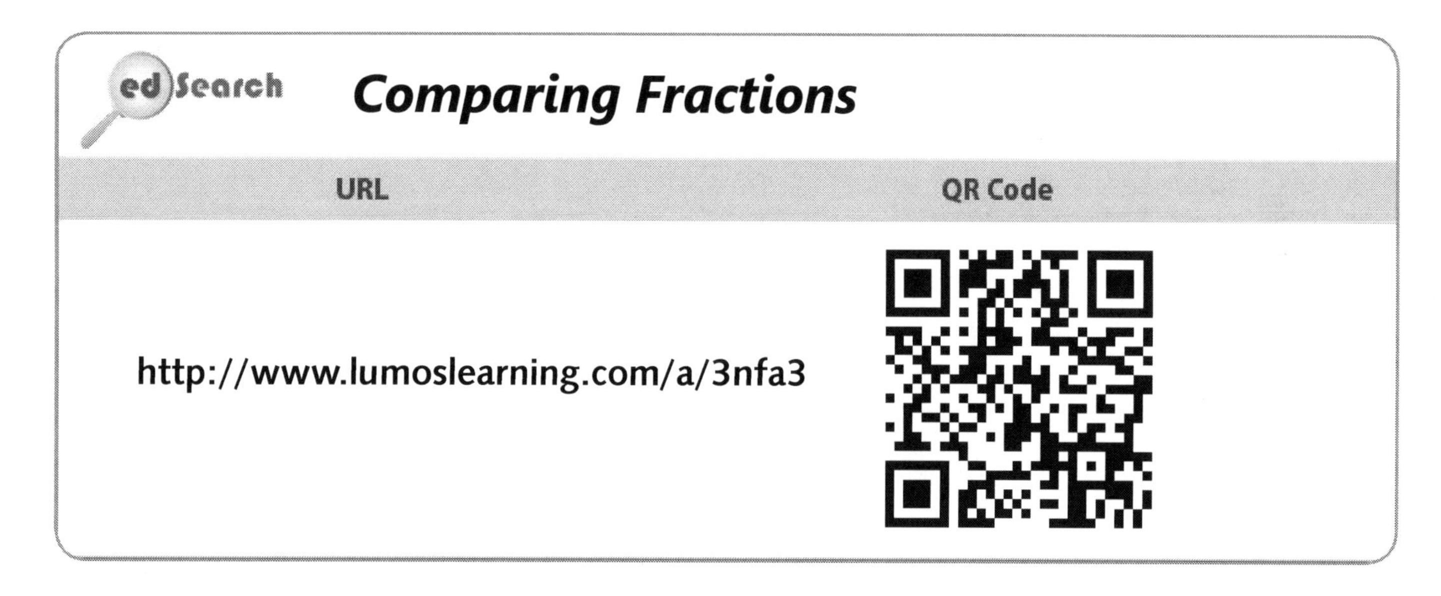

© Lumos Information Services 2018 | LumosLearning.com

1. **Which of these sets has the fractions listed from least to greatest?**

   Ⓐ $\frac{1}{6}, \frac{1}{4}, \frac{1}{3}, \frac{1}{2}$

   Ⓑ $\frac{1}{2}, \frac{1}{3}, \frac{1}{6}, \frac{1}{4}$

   Ⓒ $\frac{1}{3}, \frac{1}{4}, \frac{1}{2}, \frac{1}{6}$

   Ⓓ $\frac{1}{2}, \frac{1}{3}, \frac{1}{4}, \frac{1}{6}$

2. **Which of these fractions would be found between $\frac{1}{2}$ and 1 on a number line?**

   Ⓐ $\frac{1}{4}$

   Ⓑ $\frac{1}{3}$

   Ⓒ $\frac{5}{8}$

   Ⓓ $\frac{3}{1}$

3. **Answer the following: $\frac{6}{8} = ?$**

   Ⓐ $\frac{3}{4}$

   Ⓑ $\frac{1}{2}$

   Ⓒ $\frac{1}{4}$

   Ⓓ $\frac{8}{6}$

© Lumos Information Services 2018 | LumosLearning.com

4. Answer the following: $\frac{4}{8} = ?$

Ⓐ $\frac{1}{8}$

Ⓑ $\frac{1}{2}$

Ⓒ $\frac{1}{3}$

Ⓓ $\frac{3}{4}$

5. Answer the following: $\frac{6}{6} = ?$

Ⓐ $\frac{1}{3}$

Ⓑ $\frac{6}{1}$

Ⓒ $\frac{1}{5}$

Ⓓ $\frac{4}{4}$

6. Answer the following: $\frac{1}{2} >$ _______?

Ⓐ $\frac{1}{4}$

Ⓑ $\frac{2}{3}$

Ⓒ $\frac{4}{8}$

Ⓓ $\frac{2}{2}$

7. Which is greater: $\frac{4}{8}$ or $\frac{1}{2}$?

Ⓐ $\frac{1}{2}$

Ⓑ $\frac{4}{8}$

Ⓒ They are equal.

Ⓓ There is not enough information given.

© Lumos Information Services 2018 | LumosLearning.com

Name ______________________ Date ______________

8. Which fraction is less: $\frac{1}{4}$ or $\frac{1}{8}$?

Ⓐ $\frac{1}{4}$

Ⓑ $\frac{1}{8}$

Ⓒ They are equal.

Ⓓ There is not enough information given.

9. Which fraction is less $\frac{4}{6}$ or $\frac{1}{6}$?

Ⓐ $\frac{4}{6}$

Ⓑ $\frac{1}{6}$

Ⓒ They are equal.

Ⓓ There is not enough information given.

10. If two fractions have the same denominator, the one with a(n) __________ numerator is the greater fraction

Ⓐ smaller
Ⓑ greater
Ⓒ even
Ⓓ 4 cupcakes

11. Complete this number sentence: $\frac{4}{6}$ _____ $\frac{5}{6}$

Ⓐ >
Ⓑ <
Ⓒ =
Ⓓ There is not enough information given.

12. Complete this number sentence: $\frac{3}{8}$ _____ $\frac{5}{8}$

Ⓐ >
Ⓑ <
Ⓒ =
Ⓓ There is not enough information given.

© Lumos Information Services 2018 | LumosLearning.com

Name ______________________________ Date ________________

13. Complete this number sentence: $\frac{2}{4}$ _____ $\frac{1}{2}$

Ⓐ >
Ⓑ <
Ⓒ =
Ⓓ There is not enough information given.

14. Complete this number sentence: $\frac{1}{2}$ _____ $\frac{1}{3}$

Ⓐ >
Ⓑ <
Ⓒ =
Ⓓ There is not enough information given.

15. To compare fractions with the same numerator, you need to look at the __________.

Ⓐ numerators
Ⓑ denominators
Ⓒ factors of the numerator
Ⓓ multiples of the denominator

End of Number & Operations - Fractions

© Lumos Information Services 2018 | LumosLearning.com

Name ______________________________ Date ________________

Chapter 3:

Number & Operations - Fractions

Answer Key
&
Detailed Explanations

© Lumos Information Services 2018 | LumosLearning.com

Name ______________________ Date ____________

Lesson 1: Fractions of a Whole

Question No.	Answer	Detailed Explanation
1	C	When forming a fraction, the numerator will be the part of the whole and the denominator will be the whole or all parts together. In this case, there are 3 vowels (the part) and there are 7 total letters (the whole). The fraction should be 3/7.
2	B	When forming a fraction, the numerator will be the part of the whole and the denominator will be the whole or all parts together. In this case, there are 2 yellow tiles (the part) and there are 10 total tiles (the whole). The fraction should be 2/10.
3	C	When forming a fraction, the numerator will be the part of the whole and the denominator will be the whole or all parts together. In this case, there is 1 piece (the part) and there are 4 pieces (the whole). The fraction should be 1/4.
4	A	When forming a fraction, the numerator will be the part of the whole and the denominator will be the whole or all parts together. In this case, there is 1 shaded part (the part) and there are 2 total parts (the whole). The fraction should be 1/2.
5	B	When forming a fraction, the numerator will be the part of the whole and the denominator will be the whole or all parts together. In this case, there is 1 shaded part (the part) and there are 4 total parts (the whole). The fraction should be 1/4.
6	D	When forming a fraction, the numerator will be the part of the whole and the denominator will be the whole or all parts together. In this case, there are 3 NOT shaded parts (the part) and there are 4 total parts (the whole). The fraction should be 3/4.
7	B	When forming a fraction, the numerator will be the part of the whole and the denominator will be the whole or all parts together. In this case, there are 2 shaded parts (the part) and there are 8 total parts (the whole). The fraction should be 2/8.
8	A	When forming a fraction, the numerator will be the part of the whole and the denominator will be the whole or all parts together. In this case, there are 6 NOT shaded parts (the part) and there are 8 total parts (the whole). The fraction should be 6/8.

© Lumos Information Services 2018 | LumosLearning.com

Question No.	Answer	Detailed Explanation
9	D	When forming a fraction, the numerator will be the part of the whole and the denominator will be the whole or all parts together. In this case, there are 7 shaded parts (the part) and there are 8 total parts (the whole). The fraction should be 7/8.
10	A	When forming a fraction, the numerator will be the part of the whole and the denominator will be the whole or all parts together. In this case, there is 1 NOT shaded part (the part) and there are 8 total parts (the whole). The fraction should be 1/8.
11	B	When forming a fraction, the numerator will be the part of the whole and the denominator will be the whole or all parts together. In this case, there is 1 shaded part (the part) and there are 3 total parts (the whole). The fraction should be 1/3.
12	C	When forming a fraction, the numerator will be the part of the whole and the denominator will be the whole or all parts together. In this case, there are 2 NOT shaded parts (the part) and there are 3 total parts (the whole). The fraction should be 2/3.
13	C	When forming a fraction, the numerator will be the part of the whole and the denominator will be the whole or all parts together. In this case, you subtract the slices eaten from the total number of slices (12-8) to find that there are 4 slices left (the part). Since there were 12 total slices (the whole), the fraction should be 4/12.
14	C	Half is equivalent to dividing a number by 2 and $20 \div 2 = 10$.
15	A	A third is equivalent to dividing a number by 3 and $24 \div 3 = 8$.

© Lumos Information Services 2018 | LumosLearning.com

Lesson 2: Fractions on the Number Line

Question No.	Answer	Detailed Explanation
1	A	The number line is divided into four segments and the dot is at the first segment of the four. The fraction is 1/4.
2	A	The number line is divided into two segments and the dot is at the first segment of the two. The fraction is 1/2.
3	C	The number line is divided into eight segments and the dot is at the third segment of the eight. The fraction is 3/8.
4	B	The number line is divided into eight segments and the dot is at the sixth segment of the eight. The fraction is 6/8.
5	A	The number line is divided into six segments and the dot is at the first segment of the six. The fraction is 1/6.
6	C	The number line is divided into three segments and the dot is at the first segment of the three. The fraction is 1/3.
7	D	The number line is divided into six segments and the dot is at the second segment of the six. The fraction is 2/6.
8	A	The number line is divided into eight segments and the dot is at the fourth segment of the eight. The fraction is 4/8.
9	D	The number line is divided into eight segments and the dot is at the first segment of the eight. The fraction is 1/8.
10	B	The number line is divided into six segments and the dot is at the fifth segment of the six. The fraction is 5/6.
11	A	The number line is divided into three segments and the dot is at the second segment of the three. The fraction is 2/3.
12	B	The number line is divided into four segments and the dot is at the second segment of the four. The fraction is 2/4.
13	B	The number line is divided into eight segments and the dot is at the seventh segment of the eight. The fraction is 7/8.
14	D	The number line is divided into four segments and the dot is at the third segment of the four. The fraction is 3/4.
15	C	The number line is divided into eight segments and the dot is at the fifth segment of the eight. The fraction is 5/8.

© Lumos Information Services 2018 | LumosLearning.com

Name ______________________ Date ______________

Lesson 3: Comparing Fractions

Question No.	Answer	Detailed Explanation
1	A	If the numerators of two or more fractions are the same, the fraction with the greatest denominator is the smallest fraction. Option A is the only choice that has the numbers lined up from the greatest to the smallest denominator.
2	C	In order for a fraction to be between 1/2 and 1, it would have to be greater than 1/2. Option C is the only choice that fits this criteria. Since 4/8 = 1/2, 5/8 would be greater than 1/2. 3/1 is greater than 1 whole, so that is too large. 1/3 is smaller than 1/2.
3	A	The fraction 6/8 would be at the same point on a number line as 3/4. When simplified, 6/8 can be reduced to 3/4 by dividing both the numerator and denominator by 2.
4	B	The fraction 4/8 is at the same point on a number line as 1/2. When simplified, 4/8 can be reduced to 1/2 by dividing both the numerator and denominator by 4.
5	D	The fraction 6/6 is at the same point on a number line as 4/4. Fractions with the same numerator and denominator are always equal to 1 whole.
6	A	The answer has to be less than 1/2. Option A is the only choice that fits this criteria. 1/4 is less than 1/2, since fourths are smaller parts of a whole than halves. 4/8 is equal to 1/2, and 3/4 is greater.
7	C	The fraction 4/8 is at the same point on a number line as 1/2; therefore the two fractions are equal. When the numerator and denominator of 4/8 are each divided by 4, the fraction can be simplified to 1/2.
8	B	If the numerators of two fractions are the same, the fraction with the greater denominator is the smaller fraction. In this case, 1/8 is the smaller fraction.
9	B	When comparing fractions and the denominators are the same, compare the numerators. In this case, 1 is less than 4. In this case 1/4 is the smaller fraction.
10	B	When comparing fractions, if the denominators are the same, compare the numerators to see which one is the largest fractions. When comparing 2/9 and 4/9, 4/9 would be greater because it has a greater numerator.

© Lumos Information Services 2018 | LumosLearning.com

Question No.	Answer	Detailed Explanation
11	B	When comparing fractions where the denominators are the same, compare the numerators. In this case, 4 is less than 5. So 4/6 < 5/6 or 4/6 is the smaller fraction.
12	B	When comparing fractions where the denominators are the same, compare the numerators. In this case, 3 is less than 5. So 3/8 is the smaller fraction. So 3/8 < 5/8.
13	C	The fraction 2/4 is at the same point on a number line as 1/2; therefore the two fractions are equal.
14	A	If the numerators of two fractions are the same, the fraction with the lesser denominator is the greater fraction. In this case, 1/2 is greater than 1/3.
15	B.	If the numerators of two fractions are the same, compare the denominators. The fraction with the greater denominator is the smaller fraction.

© Lumos Information Services 2018 | LumosLearning.com

Name ______________________ Date ______________

Chapter 4: Measurement and Data

Lesson 1: Telling Time

You can scan the QR code given below or use the url to access additional EdSearch resources including videos and mobile apps related to *Telling Time*.

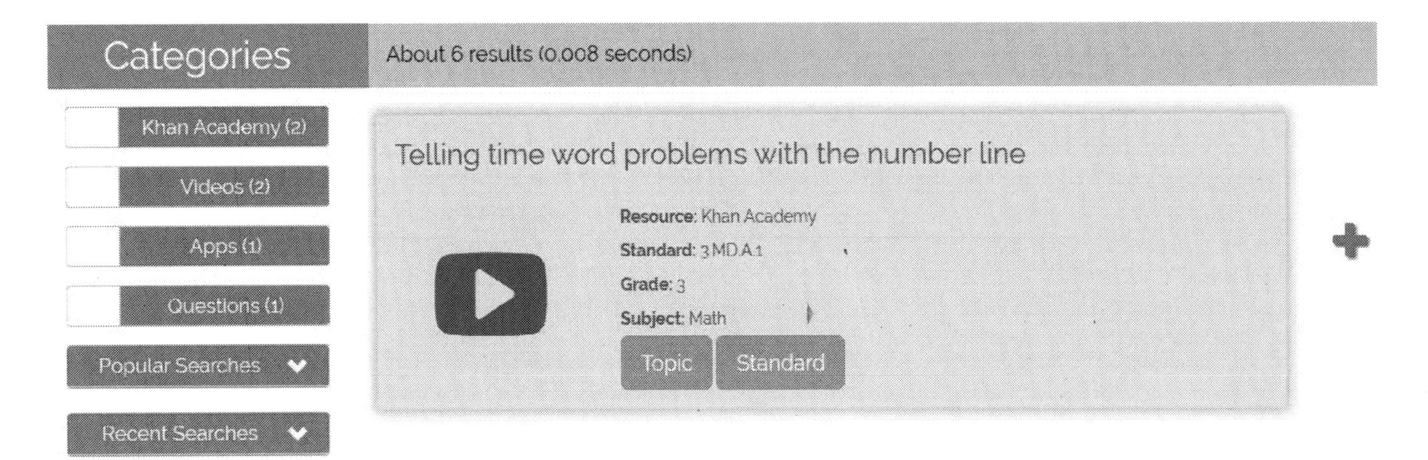

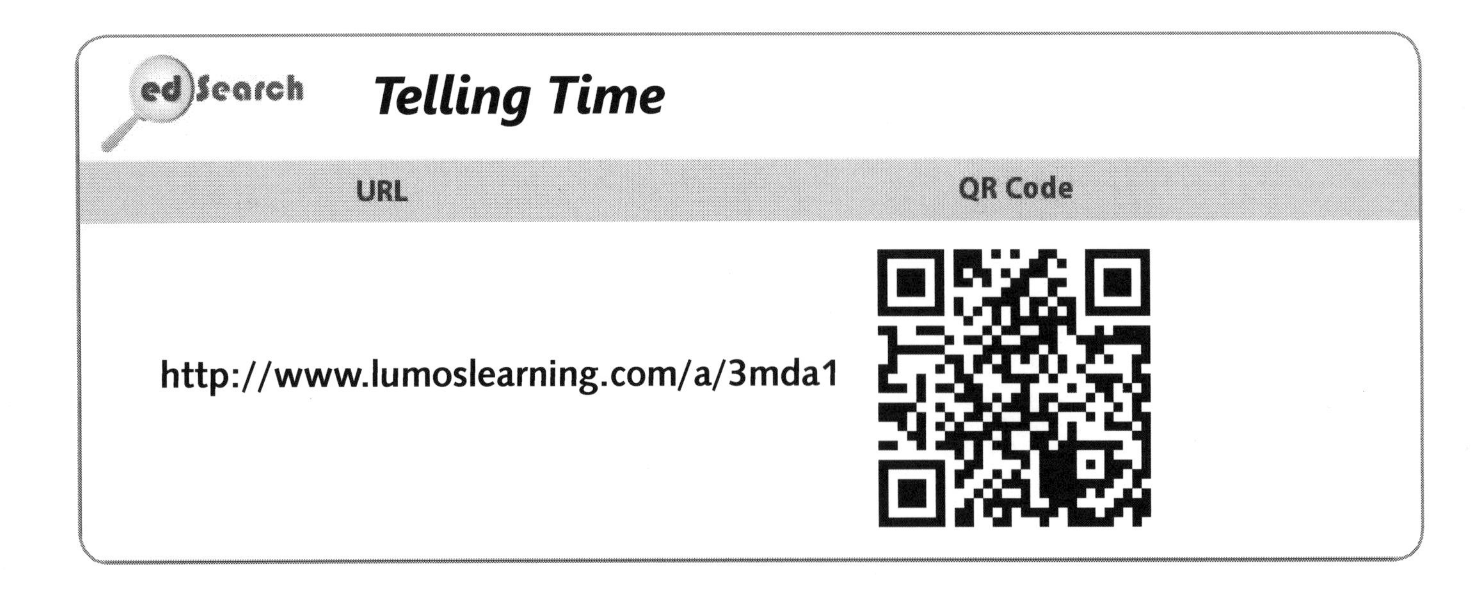

© Lumos Information Services 2018 | LumosLearning.com

1. **What time does this clock show?**

Ⓐ **3:12**
Ⓑ **2:17**
Ⓒ **2:22**
Ⓓ **2:03**

2. **What time does this clock show?**

Ⓐ **5:42**
Ⓑ **9:28**
Ⓒ **6:47**
Ⓓ **5:47**

3. **What time does this clock show?**

Ⓐ **10:00**
Ⓑ **12:50**
Ⓒ **10:02**
Ⓓ **9:41**

© Lumos Information Services 2018 | LumosLearning.com

4. What time does this clock show?

Ⓐ 12:39
Ⓑ 8:04
Ⓒ 1:38
Ⓓ 12:42

5. On an analog clock, the shorter hand shows the _____ .

Ⓐ minutes
Ⓑ hours
Ⓒ seconds
Ⓓ days

6. On an analog clock, the longer hand shows the ______ .

Ⓐ minutes
Ⓑ hours
Ⓒ days
Ⓓ seconds

7. The clock currently shows:

What time will it be in 8 minutes?

Ⓐ 1:38
Ⓑ 10:15
Ⓒ 10:10
Ⓓ 12:58

© Lumos Information Services 2018 | LumosLearning.com

Name ______________________ Date ______________

8. The clock currently shows:

What time will it be in 20 minutes?

Ⓐ 12:59
Ⓑ 1:09
Ⓒ 2:00
Ⓓ 8:24

9. The clock says:

What time was it 10 minutes ago?

Ⓐ 1:29
Ⓑ 12:29
Ⓒ 12:49
Ⓓ 1:09

10. Lucy started her test at 12:09 PM and finished at 12:58 PM. David started at 12:15 PM and ended at 1:03 PM. Who finished in a shorter amount of time?

Ⓐ Lucy
Ⓑ David
Ⓒ They both took the same amount of time.
Ⓓ There is not enough information given.

11. The Jamisons are on a road trip that will take 5 hours and 25 minutes. They have been driving for 3 hours and 41 minutes. How much longer do they need to travel before they reach their destination?

Ⓐ 1 hour, 13 minutes
Ⓑ 2 hours, 19 minutes
Ⓒ 1 hour, 44 minutes
Ⓓ 2 hours, 7 minutes

© Lumos Information Services 2018 | LumosLearning.com

12. Rachel usually gets around 9 hours of sleep per night. She went to bed at 9:30 PM. About what time will she wake up?

Ⓐ 8:30 AM
Ⓑ 10:30 AM
Ⓒ 6:30 AM
Ⓓ 5:30 AM

13. A 45 minute long show ends at 12:20 PM. When did the show begin?

Ⓐ 1:05 PM
Ⓑ 11:35 AM
Ⓒ 11:35 PM
Ⓓ 11:45 AM

14. Mrs. James is giving her class a math test. She is allowing the students 40 minutes to finish the test. The test began at 10:22 AM. By what time must the test be finished?

Ⓐ 10:42 AM
Ⓑ 10:57 AM
Ⓒ 11:02 AM
Ⓓ 12:02 PM

15. The directions on a frozen pizza say to cook it for 25 minutes. Mr. Adams puts the frozen pizza in the oven at 5:43 PM. When will the pizza be done?

Ⓐ 6:08 PM
Ⓑ 6:18 PM
Ⓒ 6:13 PM
Ⓓ 5:58 PM

© Lumos Information Services 2018 | LumosLearning.com

Chapter 4

Lesson 2: Elapsed Time

You can scan the QR code given below or use the url to access additional EdSearch resources including videos and mobile apps related to *Elapsed Time*.

© Lumos Information Services 2018 | LumosLearning.com

Name ______________________ Date ______________

1. Cedric began reading his book at 9:12 AM. He finished at 10:02 AM. How long did it take him to read his book?

Ⓐ 50 minutes
Ⓑ 40 minutes
Ⓒ 48 minutes
Ⓓ 30 minutes

2. Samantha began eating her dinner at 7:11 PM and finished at 7:35 PM so that she could go to her room and play. How long did Samantha take to eat her dinner?

Ⓐ 34 minutes
Ⓑ 21 minutes
Ⓒ 24 minutes
Ⓓ 30 minutes

3. Tanya has after school tutoring from 3:00 PM until 3:25 PM. She began walking home at 3:31 PM and arrived at her house at 3:56 PM. How long did it take Tanya to walk home?

Ⓐ 31 minutes
Ⓑ 15 minutes
Ⓒ 56 minutes
Ⓓ 25 minutes

4. Doug loves to play video games. He started playing at 4:00 PM and did not finish until 5:27 PM. How long did Doug play video games?

Ⓐ 1 hour and 37 minutes
Ⓑ 1 hour and 27 minutes
Ⓒ 27 minutes
Ⓓ 2 hours and 27 minutes

5. Kelly has to clean her room before going to bed. She began cleaning her room at 6:12 PM. When she finished, it was 7:15 PM. How long did it take Kelly to clean her room?

Ⓐ 57 minutes
Ⓑ 53 minutes
Ⓒ 1 hour and 3 minutes
Ⓓ 1 hour and 15 minutes

© Lumos Information Services 2018 | LumosLearning.com

Name ______________________________ Date ________________

6. Holly had a busy day. She attended a play from 7:06 PM until 8:13 PM. Then she went to dinner from 8:30 to 9:30 PM. How long did Holly attend the play?

Ⓐ 57 minutes
Ⓑ 2 hours and 27 minutes
Ⓒ 46 minutes
Ⓓ 1 hour and 7 minutes

7. Cara took her little brother to the park. They arrived at 3:11 PM and played until 4:37 PM. How long did Cara and her brother play at the park?

Ⓐ 26 minutes
Ⓑ 1 hour and 26 minutes
Ⓒ 56 minutes
Ⓓ 1 hour and 37 minutes

8. Arthur ran 5 miles. He began running at 8:19 AM and finished at 9:03 AM. How long did it take Arthur to run 5 miles?

Ⓐ 44 minutes
Ⓑ 45 minutes
Ⓒ 40 minutes
Ⓓ 54 minutes

9. Mr. Daniels wanted to see how fast he could wash the dishes. He began washing at 4:17 PM and finished at 4:32 PM. How long did it take Mr. Daniels to wash the dishes?

Ⓐ 15 minutes
Ⓑ 25 minutes
Ⓒ 27 minutes
Ⓓ 32 minutes

10. Sophia took a test that started at 3:28 PM. She finished the test at 4:11 PM. How long did it take Sophia to take her test?

Ⓐ 37 minutes
Ⓑ 47 minutes
Ⓒ 33 minutes
Ⓓ 43 minutes

© Lumos Information Services 2018 | LumosLearning.com

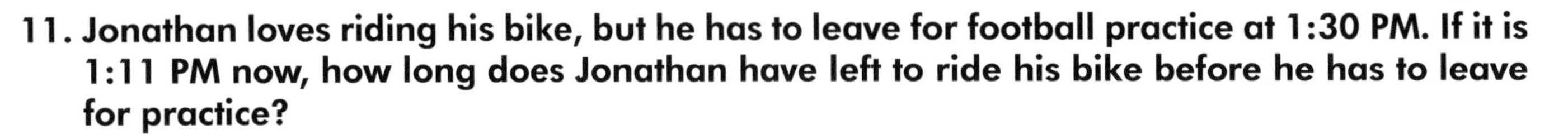

11. Jonathan loves riding his bike, but he has to leave for football practice at 1:30 PM. If it is 1:11 PM now, how long does Jonathan have left to ride his bike before he has to leave for practice?

Ⓐ 9 minutes
Ⓑ 19 minutes
Ⓒ 21 minutes
Ⓓ 29 minutes

12. Mrs. Roberts loves to take a 20-minute nap on Saturdays. She was really tired when she went to sleep at 10:45 AM. She did not wake up until 11:25 AM. How long was Mrs. Roberts' long nap?

Ⓐ 40 minutes
Ⓑ 20 minutes
Ⓒ 60 minutes
Ⓓ 30 minutes

13. Spencer has to be at his piano lesson at noon. If it is now 11:29 AM, how long does Spencer have to get to his lesson?

Ⓐ 31 minutes
Ⓑ 1 minute
Ⓒ 29 minutes
Ⓓ 39 minutes

14. Look at the clocks below. How much time has elapsed between Clock A to Clock B?

Clock A

Clock B

Ⓐ 1 hour and 2 minutes
Ⓑ 1 hour and 12 minutes
Ⓒ 52 minutes
Ⓓ 42 minutes

© Lumos Information Services 2018 | LumosLearning.com

Name ________________________________ Date __________________

15. Look at the clocks below. How much time has elapsed between Clock A to Clock B?

Clock A

Clock B

Ⓐ 52 minutes
Ⓑ 42 minutes
Ⓒ 12 minutes
Ⓓ 32 minutes

© Lumos Information Services 2018 | LumosLearning.com

Name ______________________ Date ______________

Chapter 4

Lesson 3: Liquid Volume & Mass

You can scan the QR code given below or use the url to access additional EdSearch resources including videos and mobile apps related to *Liquid Volume & Mass*.

© Lumos Information Services 2018 | LumosLearning.com

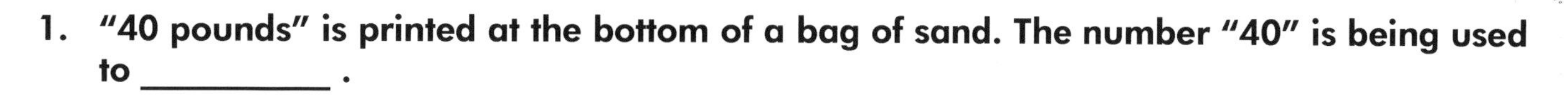

1. "40 pounds" is printed at the bottom of a bag of sand. The number "40" is being used to __________ .

Ⓐ count
Ⓑ name
Ⓒ locate
Ⓓ measure

2. In the metric system, which is the best unit to measure the mass of a coffee table?

Ⓐ Milliliters
Ⓑ Kilograms
Ⓒ Grams
Ⓓ Liters

3. Which unit should be used to measure the amount of water in a small bowl?

Ⓐ Cups
Ⓑ Gallons
Ⓒ Inches
Ⓓ Tons

4. Which unit in the customary system is best suited to measure the weight of a coffee table?

Ⓐ Gallons
Ⓑ Pounds
Ⓒ Quarts
Ⓓ Ounces

5. Which of these units could be used to measure the capacity of a container?

Ⓐ pints
Ⓑ feet
Ⓒ pounds
Ⓓ millimeters

6. Which of these is a unit of mass?

Ⓐ liter
Ⓑ meter
Ⓒ gram
Ⓓ degree

© Lumos Information Services 2018 | LumosLearning.com

Name ______________________ Date ______________

7. Which of these units has the greatest capacity?

Ⓐ gallon
Ⓑ pint
Ⓒ cup
Ⓓ quart

8. Which of these might be the weight of an average sized 8 year-old child?

Ⓐ 15 pounds
Ⓑ 30 pounds
Ⓒ 65 pounds
Ⓓ 150 pounds

9. Volume is measured in _________ units.

Ⓐ cubic
Ⓑ liters
Ⓒ square
Ⓓ box

10. What is an appropriate unit to measure the weight of a dog?

Ⓐ tons
Ⓑ pounds
Ⓒ inches
Ⓓ gallons

11. What is an appropriate unit to measure the amount of water in a swimming pool?

Ⓐ teaspoons
Ⓑ cups
Ⓒ gallons
Ⓓ inches

12. What is an appropriate unit to measure the distance across a city?

Ⓐ centimeters
Ⓑ feet
Ⓒ inches
Ⓓ miles

© Lumos Information Services 2018 | LumosLearning.com

13. What is an appropriate unit to measure the amount of salt in a cupcake recipe?

Ⓐ teaspoons
Ⓑ gallons
Ⓒ miles
Ⓓ kilograms

14. Which unit is the largest?

Ⓐ mile
Ⓑ centimeter
Ⓒ foot
Ⓓ inch

15. Which unit is the smallest?

Ⓐ kilometer
Ⓑ centimeter
Ⓒ millimeter
Ⓓ inch

Name ____________________ Date ____________

Chapter 4

Lesson 4: Graphs

You can scan the QR code given below or use the url to access additional EdSearch resources including videos and mobile apps related to *Graphs*.

edSearch ***Graphs***

URL	QR Code
http://www.lumoslearning.com/a/3mdb3	

© Lumos Information Services 2018 | LumosLearning.com

Name ______________________ Date ____________

1.

Class Survey
Should there be a field trip?

	Yes	No
Mr. A's class	𝍸 𝍸 𝍷𝍷𝍷𝍷	𝍸 𝍷𝍷
Mr. B's class	𝍸 𝍸 𝍸	𝍸 𝍸 𝍷𝍷𝍷
Mr. C's class	𝍸 𝍸 𝍷	𝍸 𝍸 𝍷
Mr. D's class	𝍸 𝍸 𝍷𝍷	𝍸 𝍸

Four 3rd grade classes in Hill Elementary School were surveyed to find out if they wanted to go on a field trip at the end of the school year. The tally table above was used to record the votes.
How many kids voted "Yes" in Mrs. B's class?

Ⓐ **28 kids**
Ⓑ **15 kids**
Ⓒ **13 kids**
Ⓓ **23 kids**

2.

Should there be a field trip?

	Yes	No
Mr. A's class	14	7
Mr. B's class	15	13
Mr. C's class	11	11
Mr. D's class	12	10
Total	52	41

Four 3rd grade classes in Hill Elementary School were surveyed to find out if they wanted to go on a field trip at the end of the school year. The table above shows the results of the survey.
How many kids voted "Yes" in Mr. A's class?

Ⓐ **7 kids**
Ⓑ **15 kids**
Ⓒ **14 kids**
Ⓓ **21 kids**

3.

Should there be a field trip?

	Yes	No
Mr. A's class	14	7
Mr. B's class	15	13
Mr. C's class	11	11
Mr. D's class	12	10
Total	52	41

© Lumos Information Services 2018 | LumosLearning.com

Name ______________________________ Date ________________

Four 3rd grade classes in Hill Elementary School were surveyed to find out if they wanted to go on a field trip at the end of the school year. The table above shows the results of the survey.
How many kids altogether voted "No" for the field trip?

Ⓐ 82 kids
Ⓑ 11 kids
Ⓒ 52 kids
Ⓓ 41 kids

4. The students in Mr. Donovan's class were surveyed to find out their favorite school subjects. The results are shown in the pictograph. Use the pictograph to answer the following question:
How many students chose either science or math?

Our Favorite Subjects

Math	OOOO
Reading	OO
Science	OOO
History	O
Other	OO

Key: O = 2 votes

Ⓐ 6 students
Ⓑ 7 students
Ⓒ 14 students
Ⓓ 2 students

5.

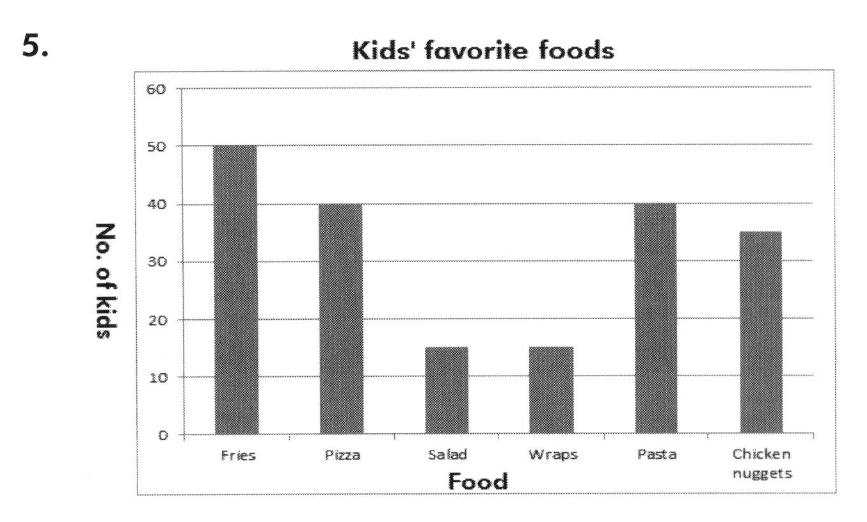

© Lumos Information Services 2018 | LumosLearning.com

Name ______________________ Date ______________

The third graders in Valley Elementary School were asked to pick their favorite food from 6 choices. The results are shown in the bar graph.
Which food was the favourite of the most children?

Ⓐ Pizza
Ⓑ Pasta
Ⓒ Fries
Ⓓ Salad

6.

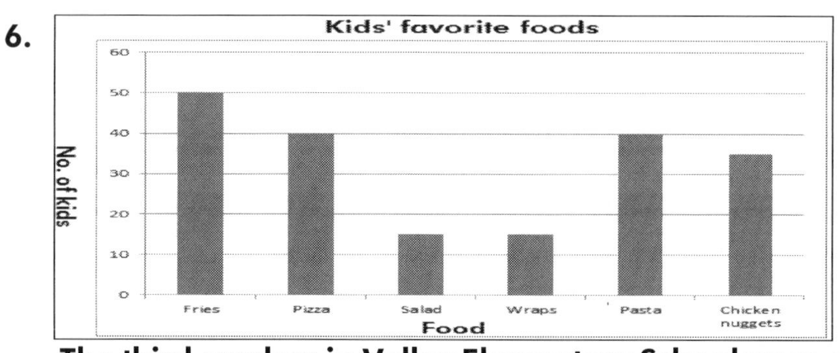

The third graders in Valley Elementary School were asked to pick their favorite food from 6 choices. The results are shown in the bar graph.

What are the 2 foods that kids like the least?

Ⓐ Fries and Pizza
Ⓑ Pizza and Pasta
Ⓒ Pasta and Chicken Nuggets
Ⓓ Salad and Wraps

7.

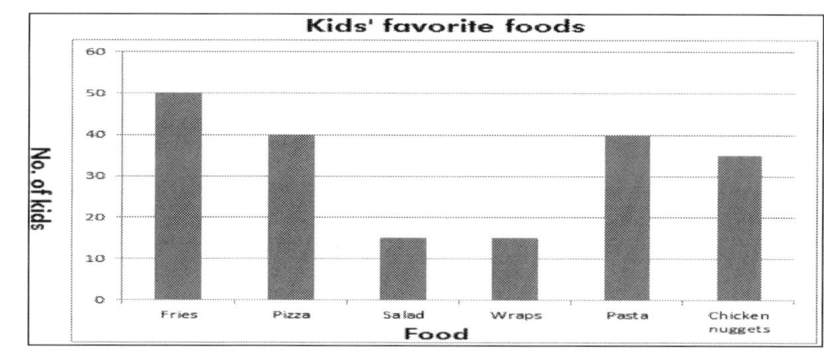

The third graders in Valley Elementary School were asked to pick their favorite food from 6 choices. The results are shown in the bar graph.
How many kids chose pasta?

Ⓐ 50 kids
Ⓑ 15 kids
Ⓒ 40 kids
Ⓓ 35 kids

© Lumos Information Services 2018 | LumosLearning.com

Name ______________________ Date ______________

8.

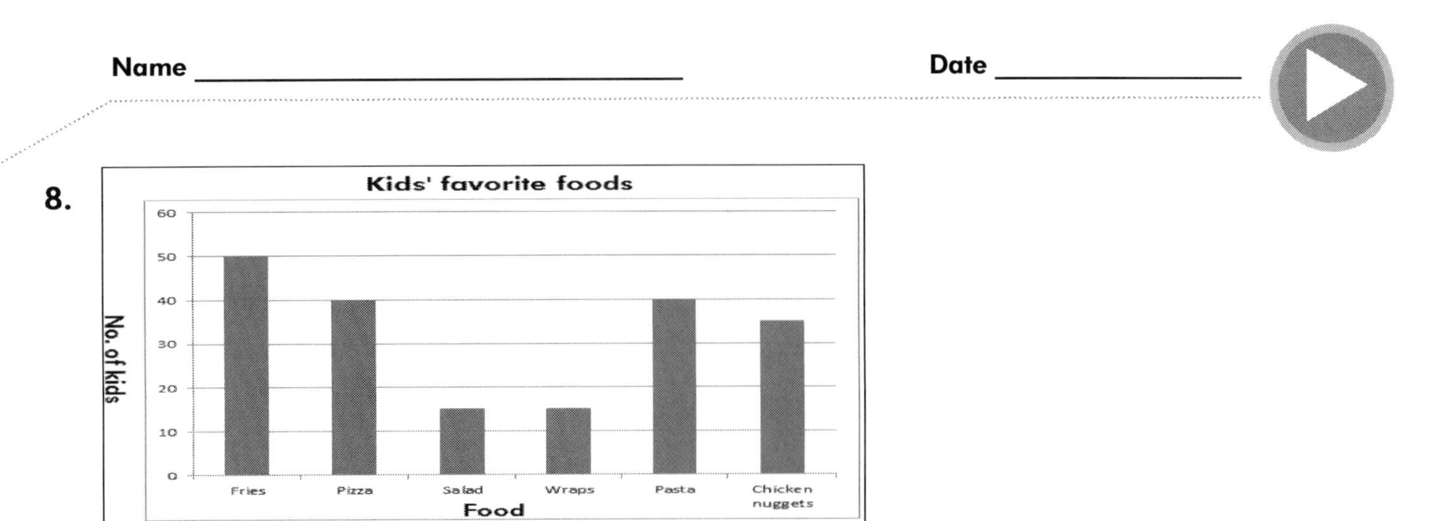

The third graders in Valley Elementary School were asked to pick their favorite food from 6 choices. The results are shown in the bar graph.

How many more kids prefer fries than pizza?

Ⓐ **50 more kids**
Ⓑ **10 more kids**
Ⓒ **1 more kid**
Ⓓ **15 more kids**

9.

Rainy Days in New Jersey

Jan	Feb	Mar	Apr	May
x	xxxx	xx	xxx	xxx

Months

Key: x = 2 days

The line plot shows the number of days it rained in New Jersey from January through May. What is the title of the above graph?

Ⓐ **Line plot**
Ⓑ **Rainy Days in New Jersey**
Ⓒ **Months**
Ⓓ **2 days**

© Lumos Information Services 2018 | LumosLearning.com

Name ______________________ Date ______________

10.

Rainy Days in New Jersey

Jan	Feb	Mar	Apr	May
x	x x x x	x x	x x x	x x x

Months

Key: x = 2 days

Which of the following statements about the above graph is true?

Ⓐ The graph shows the amount of rain accumulated each day.
Ⓑ The graph shows New Jersey's monthly rainy days from January through May.
Ⓒ The graph shows the average temperature during the 5 month period.
Ⓓ The graph shows New Jersey's total number of rainy days for the year.

11

Rainy Days in New Jersey

Jan	Feb	Mar	Apr	May
x	x x x x	x x	x x x	x x x

Months

Key: x = 2 days

According to the graph, which month had the most of rainy days?

Ⓐ March
Ⓑ February
Ⓒ January
Ⓓ April

© Lumos Information Services 2018 | LumosLearning.com

Name ______________________________ Date ________________

12. A survey was taken to find out the favorite sports of third graders in a particular class. The results are shown in the tally table. Use the table to answer the following question: How many students were surveyed altogether?

Our Favorite Sports

Soccer	~~||||~~ |
Tennis	||||
Baseball	~~||||~~ |||
Hockey	||||
Other	|||

Ⓐ 20 students
Ⓑ 25 students
Ⓒ 24 students
Ⓓ 27 students

13. A survey was taken to find out the favorite sports of third graders in a particular class. The results are shown in the tally table. Use the table to answer the following question: How many more students chose soccer than chose hockey?

Our Favorite Sports

Soccer	~~||||~~ |
Tennis	||||
Baseball	~~||||~~ |||
Hockey	||||
Other	|||

Ⓐ 6 students
Ⓑ 4 students
Ⓒ 2 students
Ⓓ 3 students

© Lumos Information Services 2018 | LumosLearning.com

14. A survey was taken to find out the favorite sports of third graders in a particular class. The results are shown in the tally table. Use the table to answer the following question: How many students chose baseball as their favorite sport?

Our Favorite Sports

Soccer	卌 |
Tennis	||||
Baseball	卌 |||
Hockey	||||
Other	|||

Ⓐ 9 students
Ⓑ 8 students
Ⓒ 3 students
Ⓓ 6 students

15. A survey was taken to find out the favorite sports of third graders in a particular class. The results are shown in the tally table. Use the table to answer the following question: Which two sports were chosen by the same number of students?

Our Favorite Sports

Soccer	卌 |
Tennis	||||
Baseball	卌 |||
Hockey	||||
Other	|||

Ⓐ soccer and tennis
Ⓑ soccer and baseball
Ⓒ hockey and soccer
Ⓓ hockey and tennis

© Lumos Information Services 2018 | LumosLearning.com

Name ____________________ Date ______________

Chapter 4

Lesson 5: Measuring Length

You can scan the QR code given below or use the url to access additional EdSearch resources including videos and mobile apps related to *Measuring Length*.

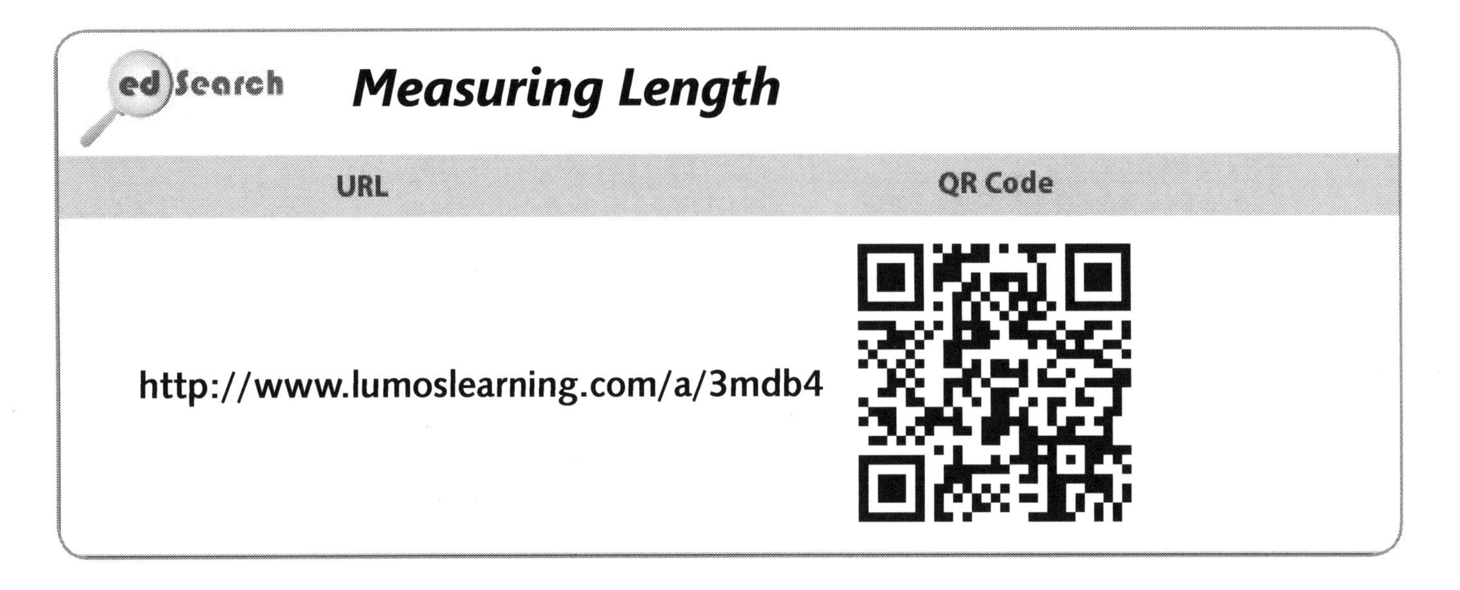

© Lumos Information Services 2018 | LumosLearning.com

Name ______________________ Date ______________

1. **Which of these units is part of the metric system?**

   Ⓐ Foot
   Ⓑ Mile
   Ⓒ Kilometer
   Ⓓ Yard

2. **Which metric unit is closest in length to one yard?**

   Ⓐ decimeter
   Ⓑ meter
   Ⓒ millimeter
   Ⓓ kilometer

3. **Which of these is the best estimate for the length of a table?**

   Ⓐ 2 decimeters
   Ⓑ 2 centimeters
   Ⓒ 2 meters
   Ⓓ 2 kilometers

4. **What unit should you use to measure the length of a book?**

   Ⓐ Kilometers
   Ⓑ Meters
   Ⓒ Centimeters
   Ⓓ Grams

5. **About how long is a new pencil?**

   Ⓐ 8 inches
   Ⓑ 8 feet
   Ⓒ 8 yards
   Ⓓ 8 miles

© Lumos Information Services 2018 | LumosLearning.com

Name ______________________ Date ______________

6. Which of these is the best estimate for the length of a football?

Ⓐ 1 foot
Ⓑ 2 feet
Ⓒ 6 feet
Ⓓ 4 feet

7. Complete the following statement.
The length of a football field is __________.

Ⓐ less than one meter
Ⓑ greater than one meter
Ⓒ about one meter
Ⓓ impossible to measure

8. Complete the following statement.
An adult's pointer finger is about one ________ wide.

Ⓐ meter
Ⓑ kilometer
Ⓒ millimeter
Ⓓ centimeter

9. Complete the following statement.
The distance between two cities would most likely be measured in ____________.

Ⓐ hours
Ⓑ miles
Ⓒ yards
Ⓓ square inches

10. A ribbon is 25 centimeters long. About how many inches long is it?

Ⓐ 2
Ⓑ 25
Ⓒ 10
Ⓓ 50

© Lumos Information Services 2018 | LumosLearning.com

11.

How long is this object?

Ⓐ 4 inches
Ⓑ 8 inches
Ⓒ 10 inches
Ⓓ 12 inches

12.

How long is this object?

Ⓐ 2 inches
Ⓑ 5 inches
Ⓒ 3 inches
Ⓓ 1 inch

13.

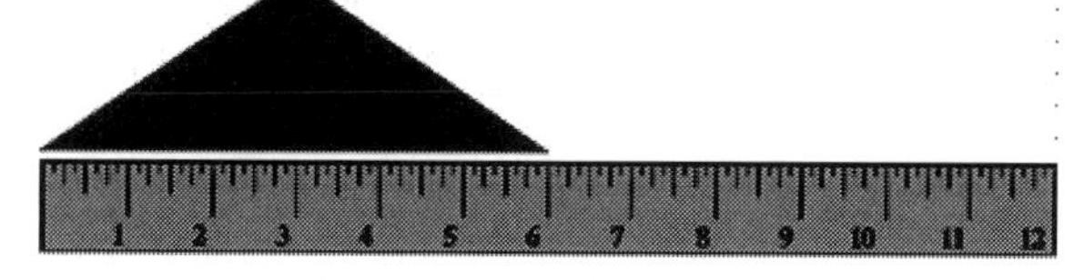

How long is this object?

Ⓐ 6 inches
Ⓑ 5 and a half inches
Ⓒ 6 and a half inches
Ⓓ 7 inches

14. Which statement is correct?

Ⓐ 1 inch > 1 mile
Ⓑ 1 inch > 1 centimeter
Ⓒ 1 foot < 1 inch
Ⓓ 1 mile < 1 foot

© Lumos Information Services 2018 | LumosLearning.com

Name ______________________ Date ______________

15. Which is correct?

Ⓐ 12 inches > 1 foot
Ⓑ 12 inches < 1 foot
Ⓒ 12 inches = 1 foot
Ⓓ 9 inches = 1 foot

© Lumos Information Services 2018 | LumosLearning.com

Chapter 4

Lesson 6: Area

You can scan the QR code given below or use the url to access additional EdSearch resources including videos and mobile apps related to *Area*.

edSearch **Area**

URL	QR Code
http://www.lumoslearning.com/a/3mdc6	

© Lumos Information Services 2018 | LumosLearning.com

Name ______________________________ Date ________________

1. The area of a plane figure is measured in __________ units.

Ⓐ cubic
Ⓑ meter
Ⓒ square
Ⓓ box

2. Which of these objects has an area of about 1 square inch?

Ⓐ a sheet of writing paper
Ⓑ a beach towel
Ⓒ a dollar bill
Ⓓ a postage stamp

3. Which of these objects would have the most square units?

Ⓐ a window with 4 panes
Ⓑ a chocolate bar with 8 squares
Ⓒ A toy block
Ⓓ A sheet of graphing paper with 24 squares

4.

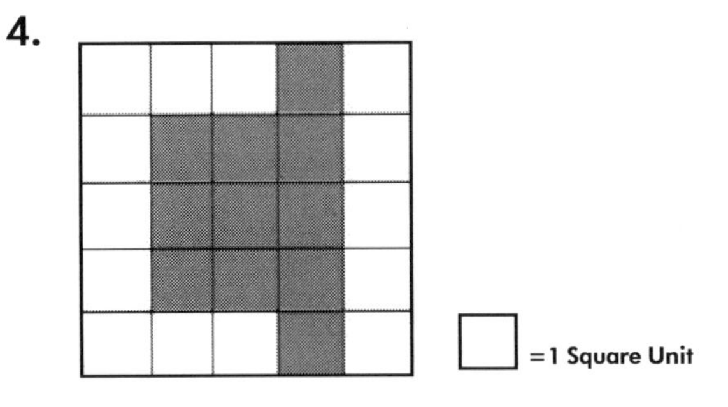

What is the area of the shaded region?

Ⓐ 10 square units
Ⓑ 8 square units
Ⓒ 11 square units
Ⓓ 15 square units

© Lumos Information Services 2018 | LumosLearning.com

5. Find the area of this figure.

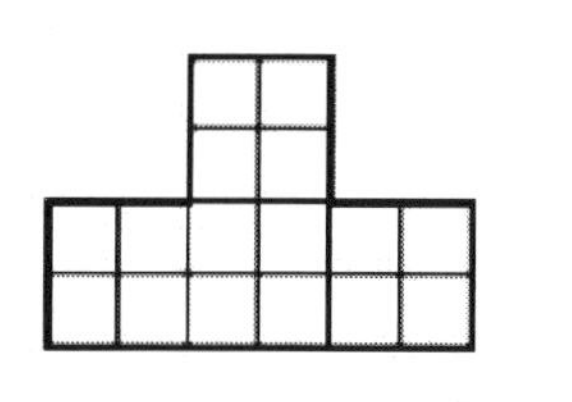

☐ =1 Square Unit

Ⓐ 22 square units
Ⓑ 20 square units
Ⓒ 18 square units
Ⓓ 16 square units

6. Find the area of this figure.

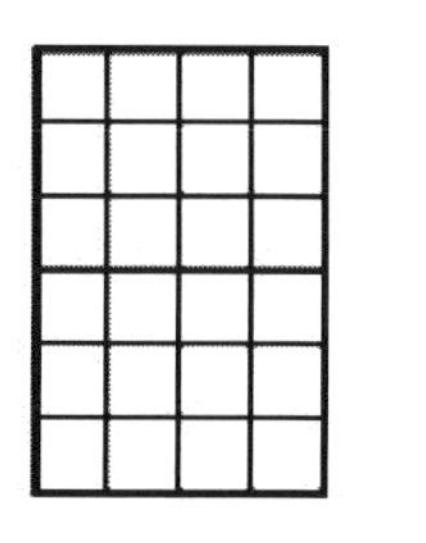

☐ =1 Square Unit

Ⓐ 22 square units
Ⓑ 20 square units
Ⓒ 24 square units
Ⓓ 28 square units

7. Find the area of the shaded region.

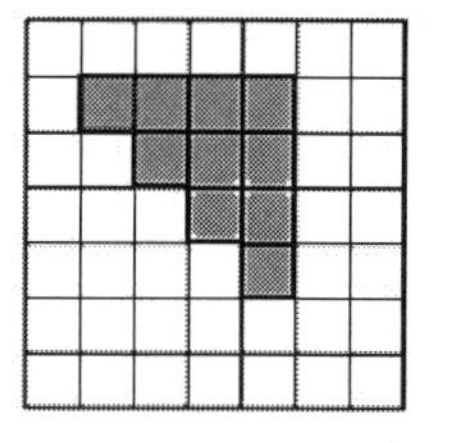

☐ =1 Square Unit

Ⓐ 11 square units
Ⓑ 10 square units
Ⓒ 16 square units
Ⓓ 9 square units

© Lumos Information Services 2018 | LumosLearning.com

8. Find the area of the shaded region.

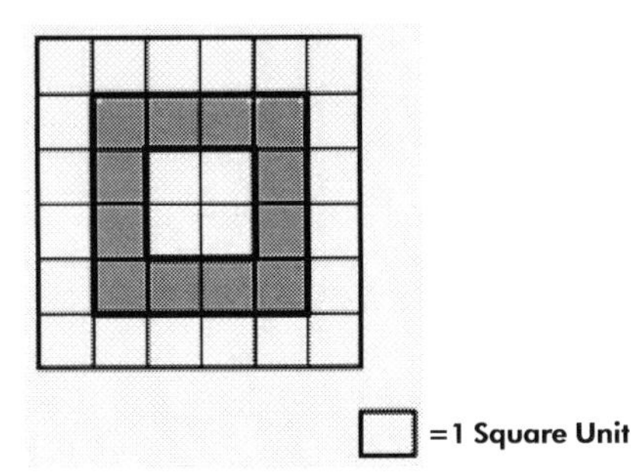

Ⓐ 16 square units
Ⓑ 12 square units
Ⓒ 10 square units
Ⓓ 11 square units

9. Find the area of the shaded region.

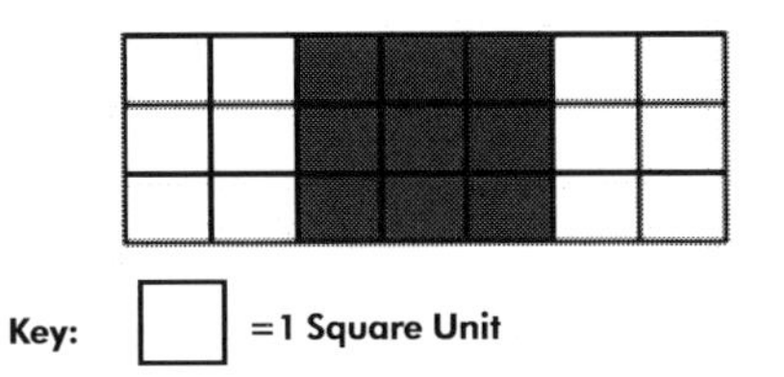

Ⓐ 5 square units
Ⓑ 6 square units
Ⓒ 8 square units
Ⓓ 9 square units

10. Find the area of the shaded region.

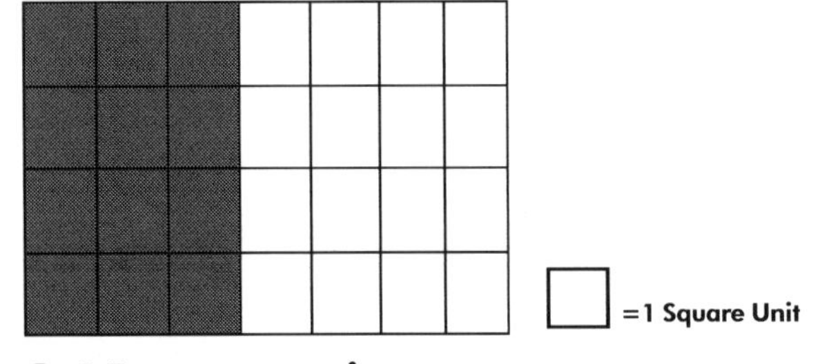

Ⓐ 15 square units
Ⓑ 28 square units
Ⓒ 12 square units
Ⓓ 7 square units

© Lumos Information Services 2018 | LumosLearning.com

11. Find the area of the shaded region.

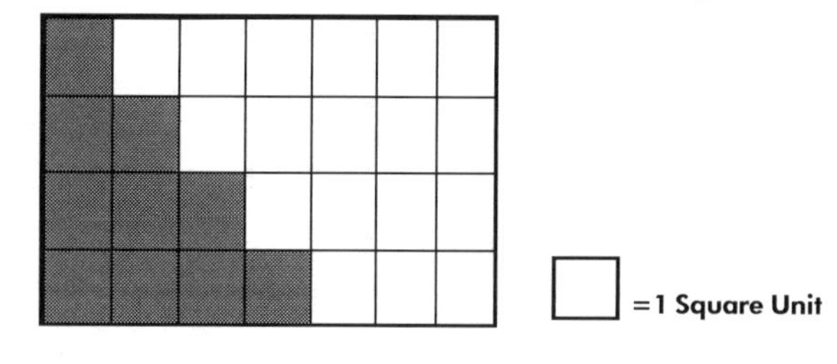

= 1 Square Unit

Ⓐ 10 square units
Ⓑ 20 square units
Ⓒ 11 square units
Ⓓ 4 square units

12. Find the area of this region.

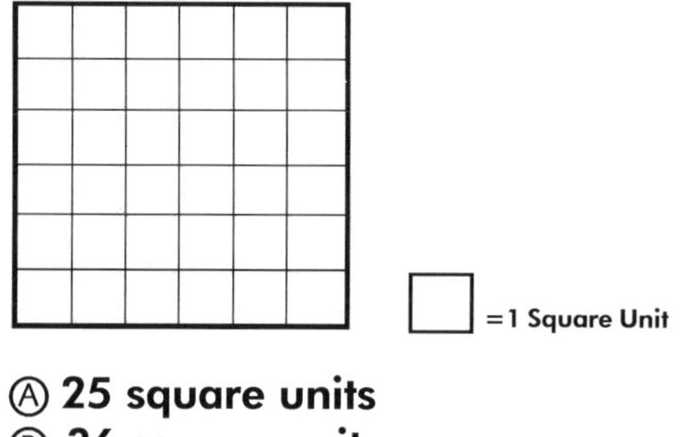

= 1 Square Unit

Ⓐ 25 square units
Ⓑ 36 square units
Ⓒ 32 square units
Ⓓ 17 square units

13. Which figure has the larger area?

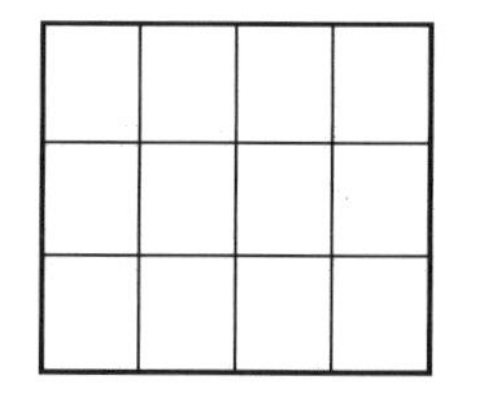

Figure A

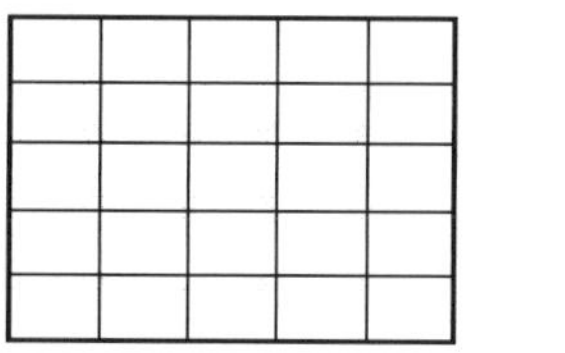

Figure B

= 1 Square Unit

Ⓐ Figure A
Ⓑ Figure B
Ⓒ Both figures have the same area
Ⓓ There is not enough information

© Lumos Information Services 2018 | LumosLearning.com

14. Which two figures have the same shaded area?

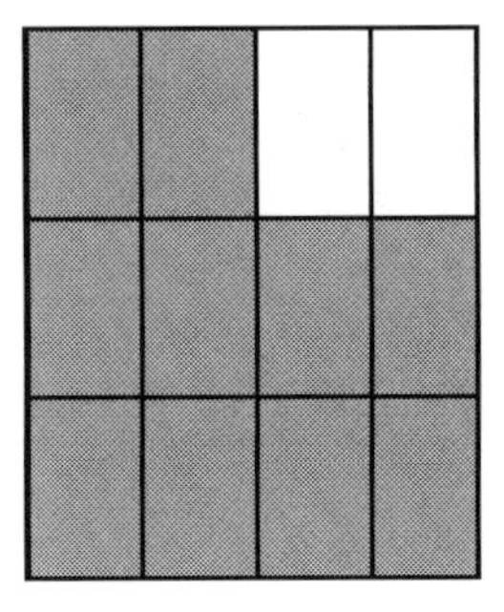

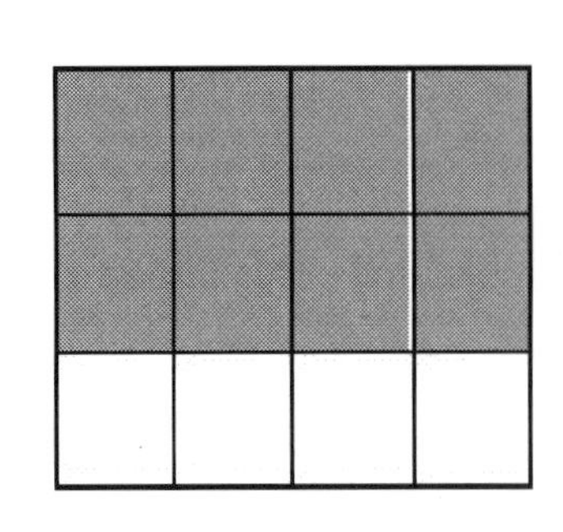

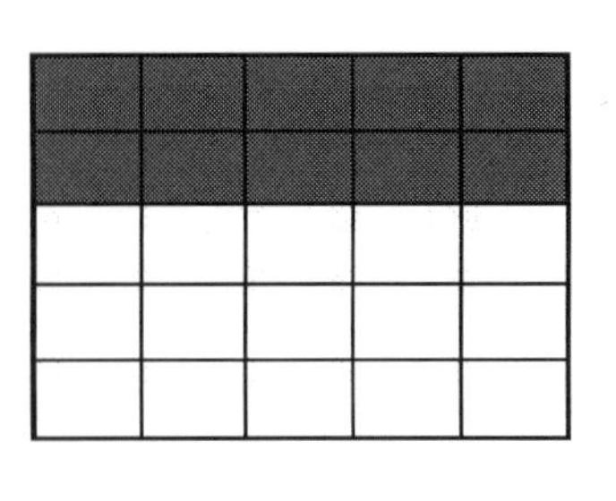

=1 Square Unit

Figure A

Figure B

Figure C

Ⓐ Figures A and B
Ⓑ Figures A and C
Ⓒ Figures B and C
Ⓓ All three figures have the same shaded area

15. Find the area of this figure.

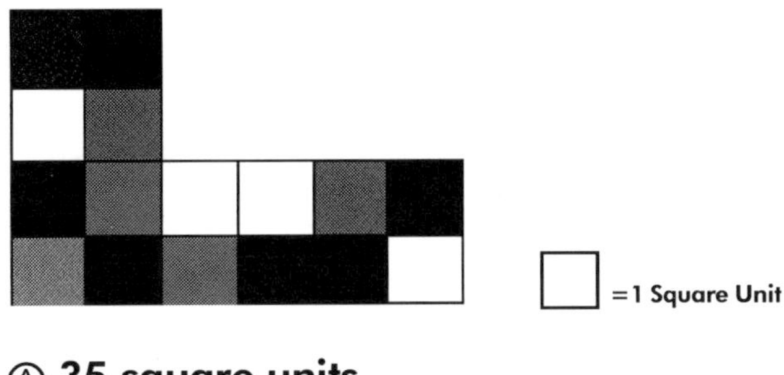

Ⓐ 35 square units
Ⓑ 26 square units
Ⓒ 22 square units
Ⓓ 16 square units

© Lumos Information Services 2018 | LumosLearning.com

Name ______________________ Date ______________

Chapter 4

Lesson 7: Relating Area to Addition & Multiplication

You can scan the QR code given below or use the url to access additional EdSearch resources including videos and mobile apps related to *Relating Area to Addition & Multiplication*.

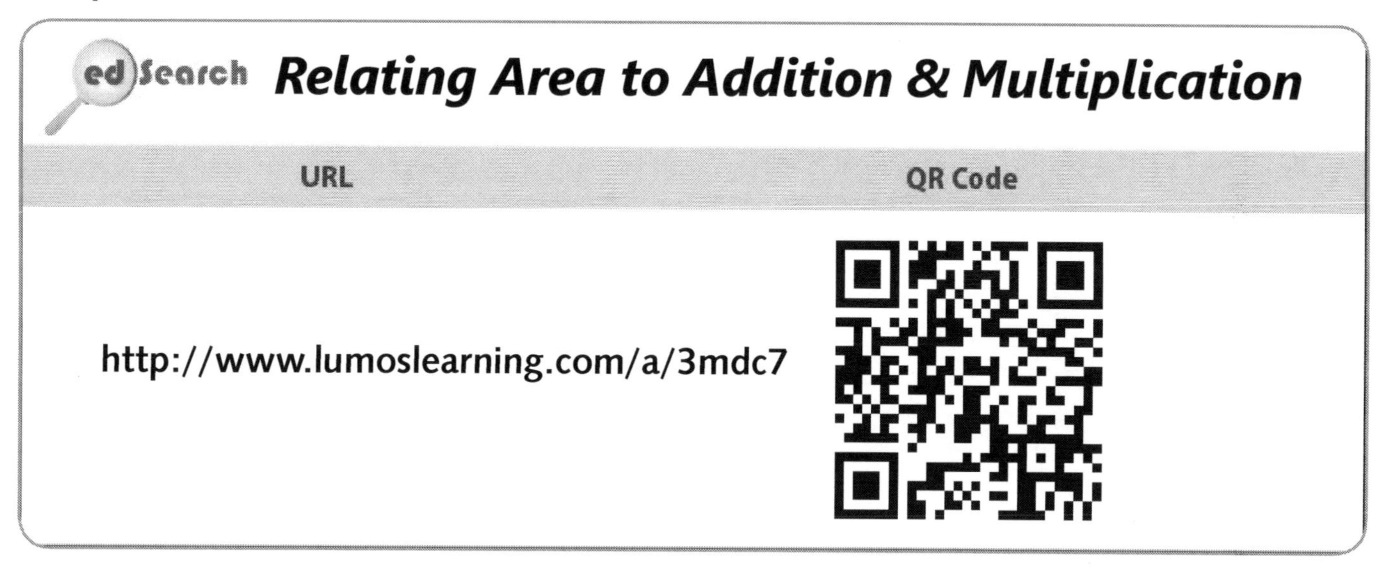

© Lumos Information Services 2018 | LumosLearning.com

Name ______________________ Date ______________

1. **Find the area of the object below.**

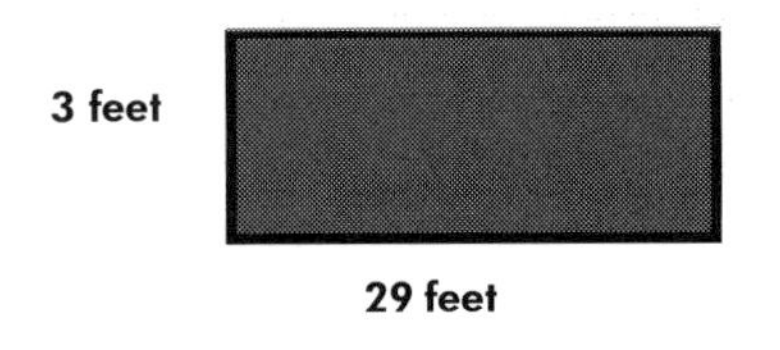

Ⓐ 87 square feet
Ⓑ 32 square feet
Ⓒ 64 square feet
Ⓓ 128 square feet

2. **Find the area of the object below.**

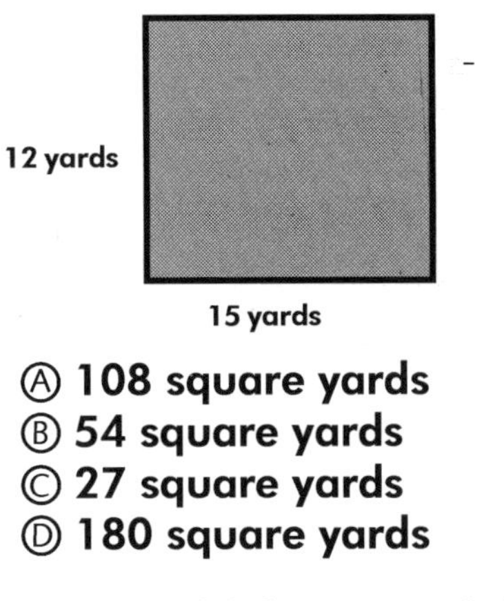

Ⓐ 108 square yards
Ⓑ 54 square yards
Ⓒ 27 square yards
Ⓓ 180 square yards

3. **How could the area of this figure be calculated?**

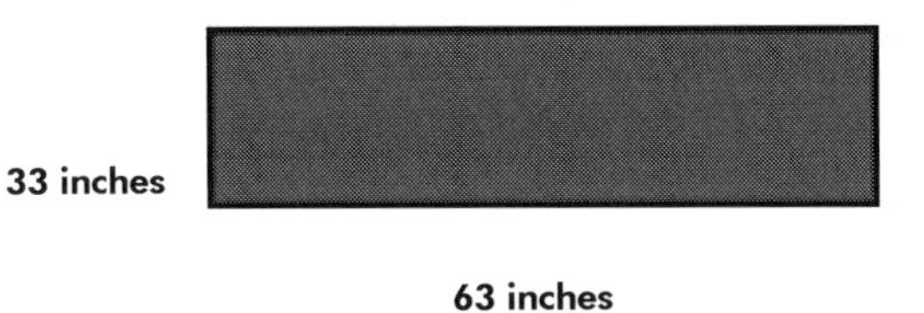

Ⓐ Multiply 63 x 33 x 63 x 33
Ⓑ Add 63 + 33 + 63 + 33
Ⓒ Multiply 63 x 33
Ⓓ Multiply 2 x 63 x 33

© Lumos Information Services 2018 | LumosLearning.com

4. **Find the area of the object below.**

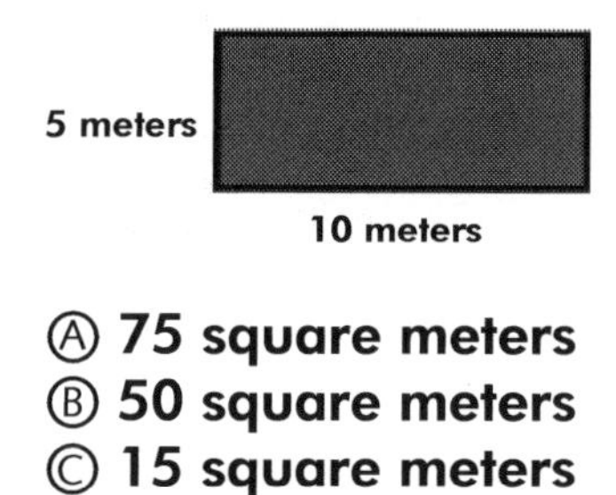

Ⓐ 75 square meters
Ⓑ 50 square meters
Ⓒ 15 square meters
Ⓓ 30 square meters

5. **Find the area of the object below.**

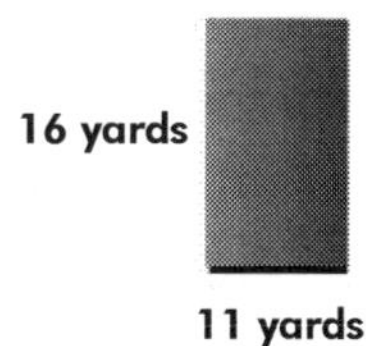

Ⓐ 176 square yards
Ⓑ 27 square yards
Ⓒ 54 square yards
Ⓓ 2,916 square yards

6. **Find the area of the object below.**

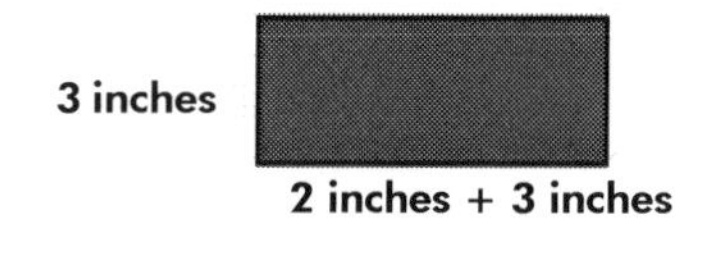

Ⓐ 18 square inches
Ⓑ 15 square inches
Ⓒ 9 square inches
Ⓓ 6 square inches

7. **Find the area of the object below.**

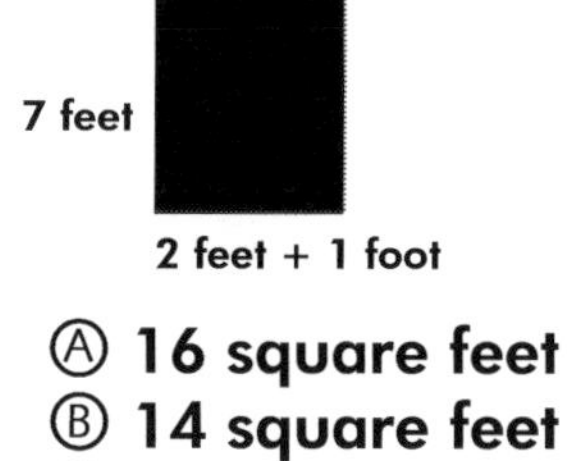

Ⓐ 16 square feet
Ⓑ 14 square feet
Ⓒ 9 square feet
Ⓓ 21 square feet

© Lumos Information Services 2018 | LumosLearning.com

Name ______________________ Date 

8. **Find the area of the object below.**

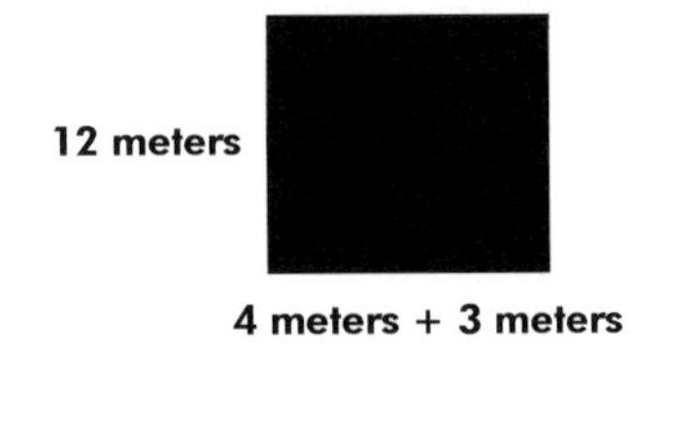

Ⓐ 84 square meters
Ⓑ 48 square meters
Ⓒ 36 square meters
Ⓓ 72 square meters

9. **Find the area of the object below.**

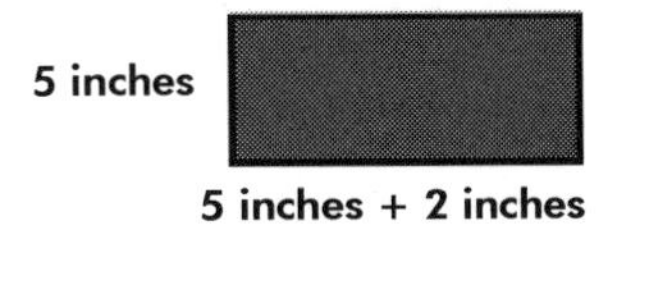

Ⓐ 12 square inches
Ⓑ 10 square inches
Ⓒ 25 square inches
Ⓓ 35 square inches

10. **Find the area of the object below.**

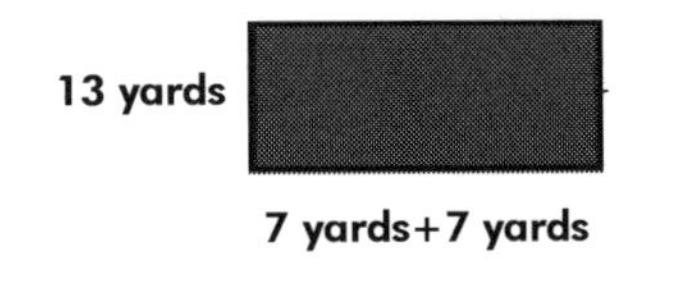

Ⓐ 84 square yards
Ⓑ 182 square yards
Ⓒ 26 square yards
Ⓓ 19 square yards

11. **The city wants to plant grass in a park. The park is 20 feet by 50 feet. How much grass will they need to cover the entire park?**

Ⓐ 100 square feet
Ⓑ 500 square feet
Ⓒ 1,000 square feet
Ⓓ 1,100 square feet

© Lumos Information Services 2018 | LumosLearning.com

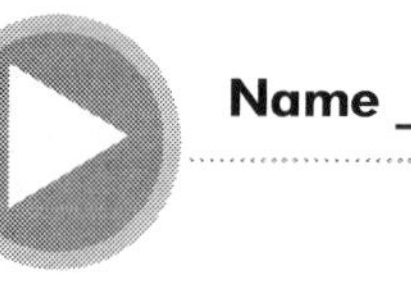

Name ______________________ Date ______________

12. Brenda wants to purchase a rug for her room. Her room is a rectangle that measures 7 yards by 6 yards. What is the area of her room?

Ⓐ 42 square yards
Ⓑ 48 square yards
Ⓒ 36 square yards
Ⓓ 26 square yards

13. Joan wants to cover her backyard with flowers. If her backyard is 30 feet long and 20 feet wide, what is the area that needs to be covered in flowers?

Ⓐ 60 square feet
Ⓑ 500 square feet
Ⓒ 600 square feet
Ⓓ 100 square feet

14. Bethany decided to paint the four walls in her room. If each wall measures 20 feet by 10 feet, how many total square feet will she need to paint?

Ⓐ 400 square feet
Ⓑ 200 square feet
Ⓒ 800 square feet
Ⓓ 600 square feet

15. Seth wants to cover his table top with a piece of fabric. His table is 2 meters long and 4 meters wide. How much fabric does Seth need?

Ⓐ 6 square meters
Ⓑ 10 square meters
Ⓒ 8 square meters
Ⓓ 16 square meters

© Lumos Information Services 2018 | LumosLearning.com

Chapter 4

Lesson 8: Perimeter

You can scan the QR code given below or use the url to access additional EdSearch resources including videos and mobile apps related to *Perimeter*.

edSearch **Perimeter**

URL	QR Code
http://www.lumoslearning.com/a/3mdd8	

© Lumos Information Services 2018 | LumosLearning.com

Name ______________________ Date ______________

1. What is meant by the "perimeter" of a shape?

Ⓐ The distance from the center of a plane figure to its edge
Ⓑ The distance from one corner of a plane figure to an opposite corner
Ⓒ The distance around the outside of a plane figure
Ⓓ The amount of space covered by a plane figure

2. Complete the following statement.
Two measurements associated with plane figures are ____________.

Ⓐ perimeter and volume
Ⓑ perimeter and area
Ⓒ volume and area
Ⓓ weight and volume

3. The _______ of a figure can be found by adding the lengths of its 4 sides.

Ⓐ diameter
Ⓑ perimeter
Ⓒ area
Ⓓ circumference

4. Which of the following formulas can be used for calculating the perimeter of a rectangle?

Ⓐ a x b=
Ⓑ a + b=
Ⓒ a + b + a + b=
Ⓓ a x b + c=

5.

This rectangle is 4 units long and one unit wide. What is its perimeter?

Ⓐ 10 units
Ⓑ 4 units
Ⓒ 5 units
Ⓓ 8 units

© Lumos Information Services 2018 | LumosLearning.com

6.

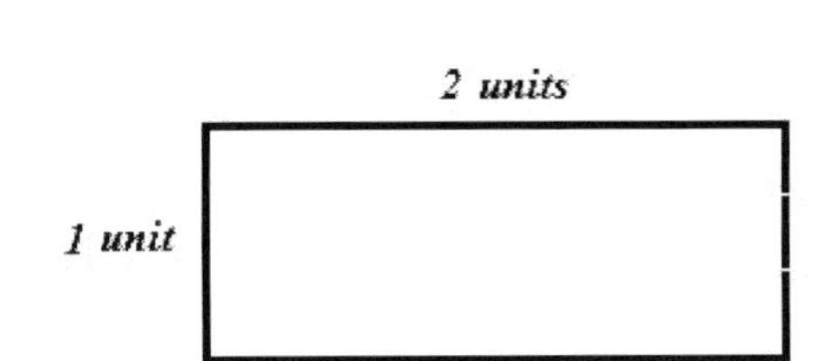

What is the perimeter of the rectangle?

Ⓐ 5 units
Ⓑ 6 units
Ⓒ 3 units
Ⓓ 2 units

7.

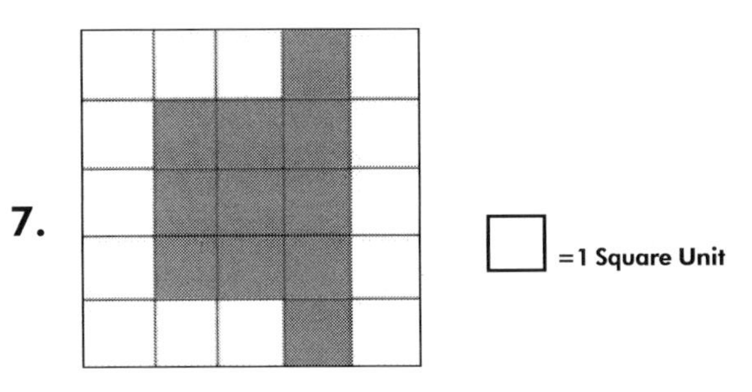

What is the perimeter of the shaded region in the above figure?

Ⓐ 16 units
Ⓑ 15 units
Ⓒ 11 units
Ⓓ 10 units

8. The perimeter of this rhombus is 20 units. How long is each of its sides?

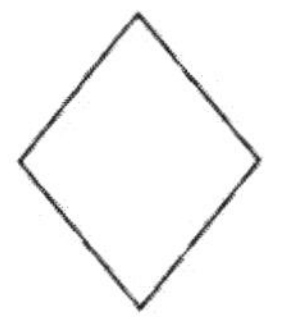

Ⓐ 4 units
Ⓑ 10 units
Ⓒ 5 units
Ⓓ This cannot be determined.

© Lumos Information Services 2018 | LumosLearning.com

9. Each side of this rhombus measures 3 centimeters. What is its perimeter?

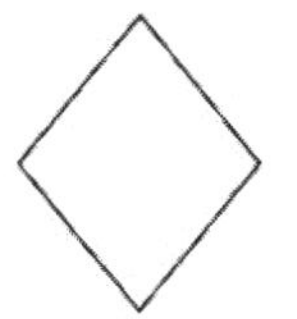

Ⓐ 3 centimeters
Ⓑ 12 centimeters
Ⓒ 9 centimeters
Ⓓ 6 centimeters

10. This square has a perimeter of 80 units. How long is each of its sides?

Ⓐ 8 units
Ⓑ 10 units
Ⓒ 20 units
Ⓓ 40 units

11. Find the perimeter of this figure.

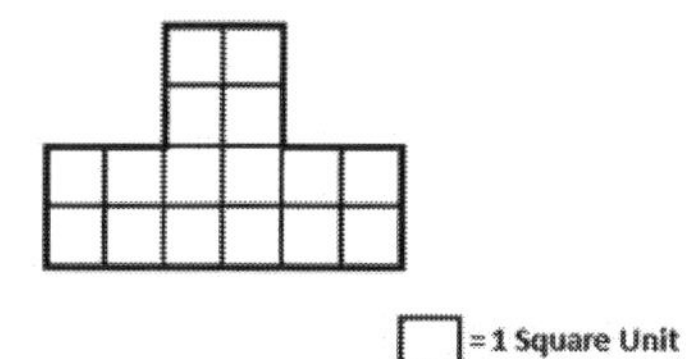

Ⓐ 20 units
Ⓑ 18 units
Ⓒ 16 units
Ⓓ 22 units

12. Joan wants to cover the outside border of her backyard with flowers. If her backyard is 30 feet long and 15 feet wide, how many feet of flowers does she need to plant?

Ⓐ 450 feet
Ⓑ 90 feet
Ⓒ 60 feet
Ⓓ 30 feet

© Lumos Information Services 2018 | LumosLearning.com

13. Find the perimeter of this figure.

= 1 Square Unit

Ⓐ 20 units
Ⓑ 18 units
Ⓒ 24 units
Ⓓ 22 units

14. Brenda wants to place rope around a large field in order to play a game. The field is a rectangle that measures 23 yards by 32 yards. How much rope does Brenda need?

Ⓐ 64 yards
Ⓑ 736 yards
Ⓒ 110 yards
Ⓓ 55 yards

15. Find the perimeter of the following object.

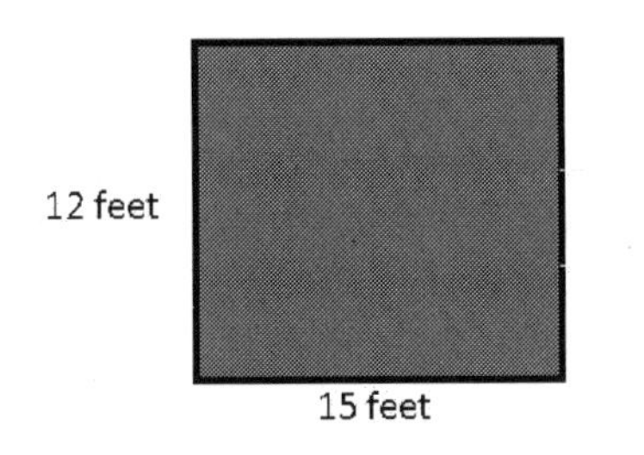

Ⓐ 54 feet
Ⓑ 27 feet
Ⓒ 58 feet
Ⓓ 180 feet

16. Find the perimeter of the following object.

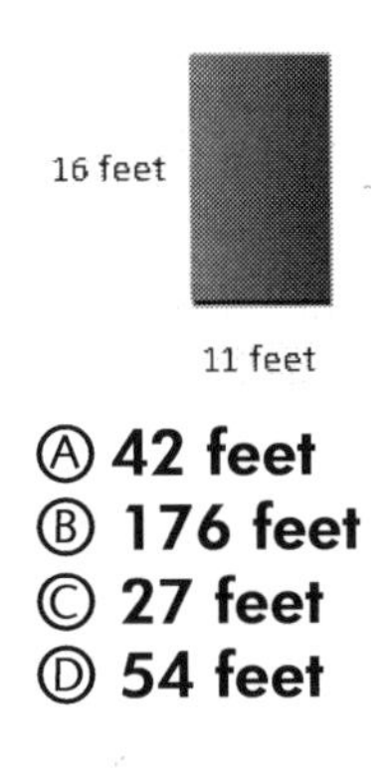

Ⓐ 42 feet
Ⓑ 176 feet
Ⓒ 27 feet
Ⓓ 54 feet

© Lumos Information Services 2018 | LumosLearning.com

17. **Find the perimeter of the shaded region.**

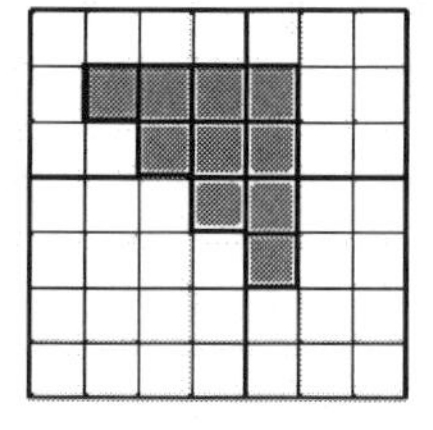

= 1 Square Unit

Ⓐ **10 units**
Ⓑ **13 units**
Ⓒ **15 units**
Ⓓ **16 units**

18. **Find the perimeter of the following object.**

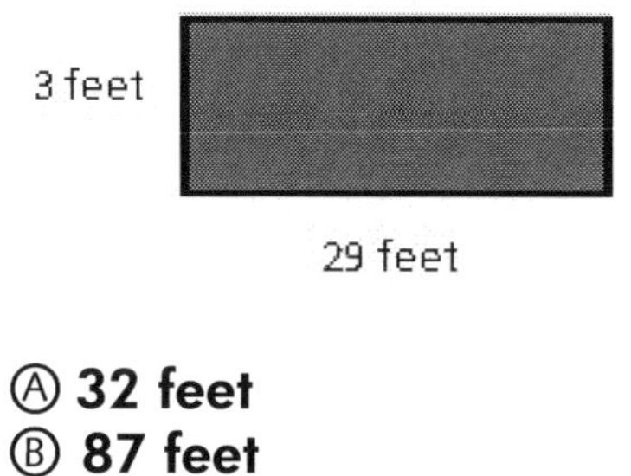

Ⓐ **32 feet**
Ⓑ **87 feet**
Ⓒ **172 feet**
Ⓓ **64 feet**

19. **The city is building a fence around a park. The park is 20 feet by 50 feet. How many feet of fencing do they need?**

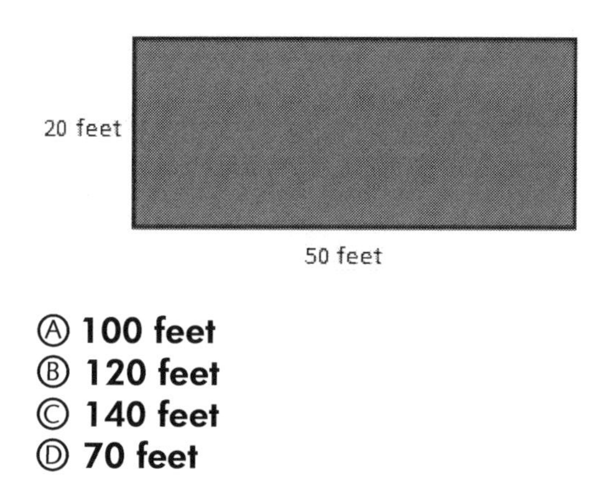

Ⓐ **100 feet**
Ⓑ **120 feet**
Ⓒ **140 feet**
Ⓓ **70 feet**

© Lumos Information Services 2018 | LumosLearning.com

Name ______________________________ Date ________________

20. Find the perimeter of the following object.

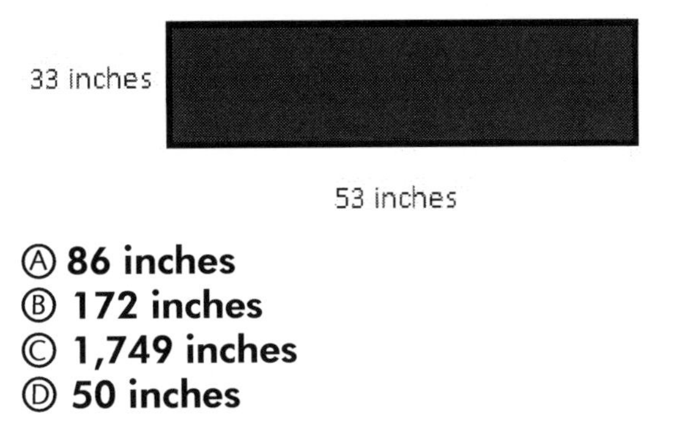

Ⓐ 86 inches
Ⓑ 172 inches
Ⓒ 1,749 inches
Ⓓ 50 inches

21. Find the perimeter of the following object.

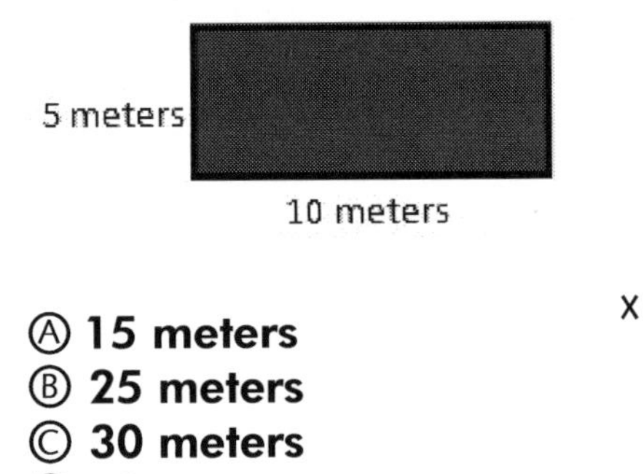

Ⓐ 15 meters
Ⓑ 25 meters
Ⓒ 30 meters
Ⓓ 50 meters

x

22. The perimeter of the following object is 16 feet. Find the length of the missing side.

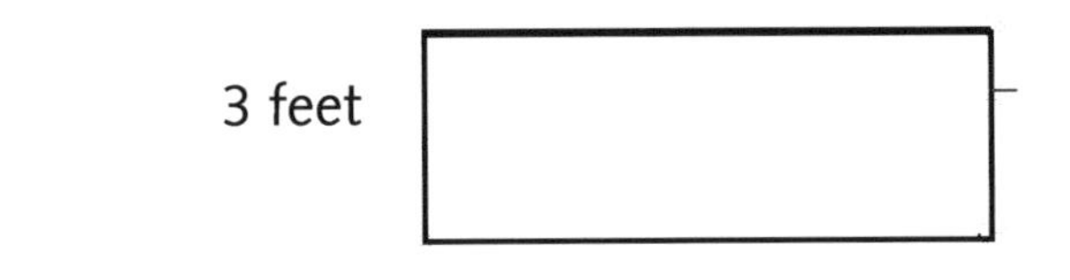

Ⓐ x = 5 feet
Ⓑ x = 10 feet
Ⓒ x = 13 feet
Ⓓ x = 6 feet

© Lumos Information Services 2018 | LumosLearning.com

23. The perimeter of the following object is 20 feet. Find the length of the missing side.

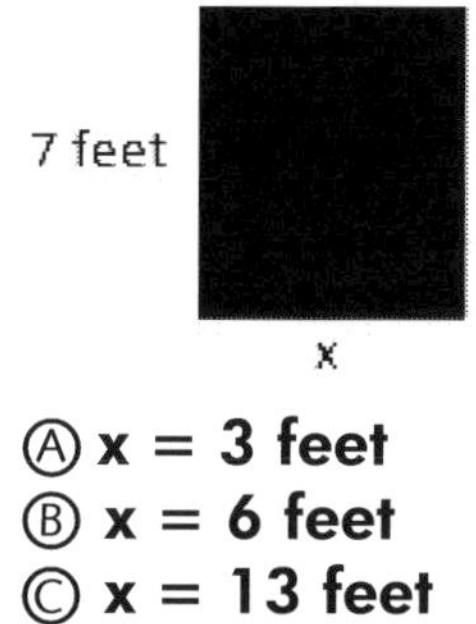

Ⓐ x = 3 feet
Ⓑ x = 6 feet
Ⓒ x = 13 feet
Ⓓ x = 7 feet

24. The city is building a fence around a park. The park is 20 feet by 50 feet. If they only want the fence on 3 sides, what is the least amount of fencing they could buy?

Ⓐ 140 feet
Ⓑ 100 feet
Ⓒ 120 feet
Ⓓ 90 feet

25. The perimeter of the following object is 38 feet. Find the length of the missing side.

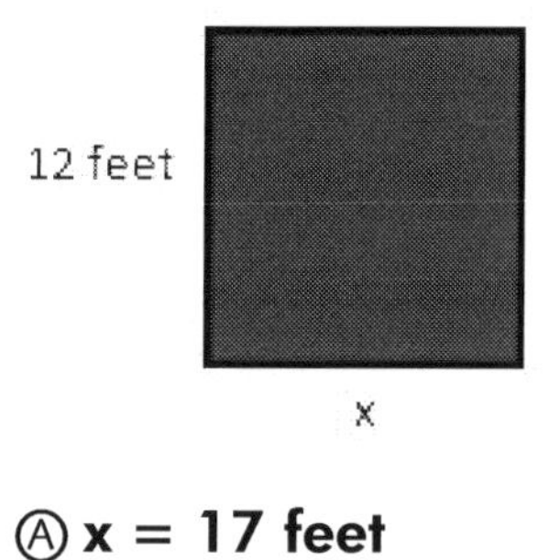

Ⓐ x = 17 feet
Ⓑ x = 12 feet
Ⓒ x = 7 feet
Ⓓ x = 26 feet

26. The perimeter of the following object is 24 inches. Find the length of the missing side.

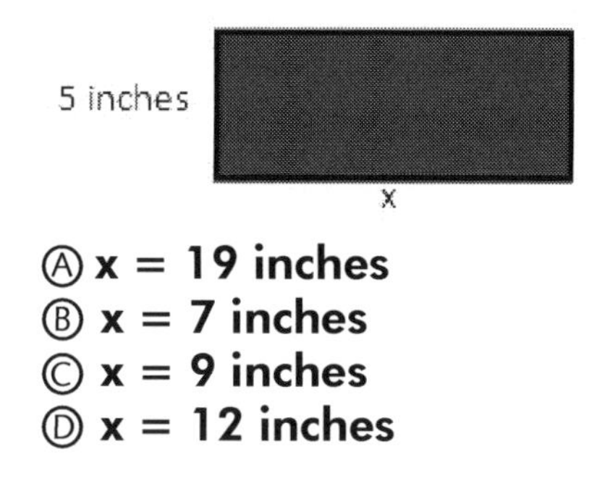

Ⓐ x = 19 inches
Ⓑ x = 7 inches
Ⓒ x = 9 inches
Ⓓ x = 12 inches

© Lumos Information Services 2018 | LumosLearning.com

Name ______________________ Date ______________

27. The perimeter of the following object is 54 yards. Find the length of the missing side.

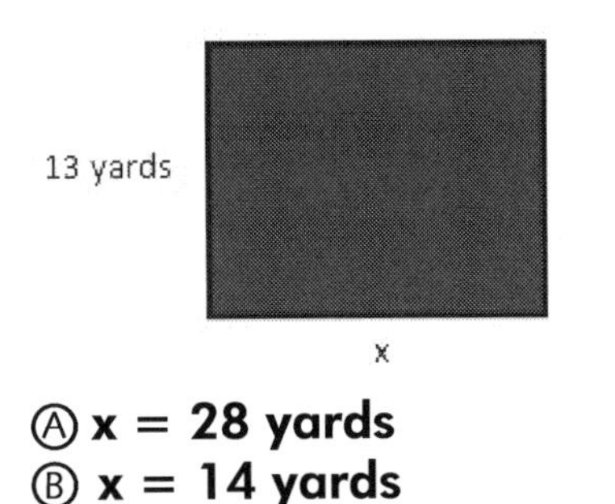

Ⓐ x = 28 yards
Ⓑ x = 14 yards
Ⓒ x = 41 yards
Ⓓ x = 27 yards

28. The perimeter of Lisa's cake is 44 inches. What size cake pan could she have used to bake her cake?

Ⓐ 13" x 9" pan
Ⓑ 8" x 8" pan
Ⓒ 12" x 6" pan
Ⓓ 9" round pan

29. Which of the following figures could have a perimeter of 44 inches?

Ⓐ A rectangle with two 7 inch sides and two 15 inch sides
Ⓑ A rectangle with two 6 inch sides and two 11 inch sides
Ⓒ A rectangle with two 14 inch sides and two 22 inch sides
Ⓓ A rectangle with two 9 inch sides and two 18 inch sides

30. Which of the following figures could have a perimeter of 56 inches?

Ⓐ A rectangle with two 3 inch sides and two 16 inch sides
Ⓑ A rectangle with two 16 inch sides and two 7 inch sides
Ⓒ A rectangle with two 20 inch sides and two 8 inch sides
Ⓓ A rectangle with two 5 inch sides and two 10 inch sides

End of Measurement & Data

© Lumos Information Services 2018 | LumosLearning.com

Chapter 4:

Measurement & Data

Answer Key
&
Detailed Explanations

© Lumos Information Services 2018 | LumosLearning.com

Lesson 1: Telling Time

Question No.	Answer	Detailed Explanation
1	B	The hour hand (the shorter hand) is past the 2nd hour but has not reached the 3rd hour, and the minute hand (the longer hand) is past 15 minutes but has not yet reached 20 minutes.
2	D	The hour hand (the shorter hand) is past the 5th hour but has not reached the 6th hour, and the minute hand (the longer hand) is past 45 minutes but has not yet reached 50 minutes.
3	C	The hour hand (the shorter hand) is pointing to the 10th hour, and the minute hand (the longer hand) is past 0 minutes but has not yet reached 5 minutes.
4	A	The hour hand (the shorter hand) is past the 12th hour but has not reached the 1st hour, and the minute hand (the longer hand) is past 35 minutes but has not yet quite reached 40 minutes.
5	B	On a clock, the shorter hand points toward the hour and the longer hand points toward the minutes. For example, if it was 2:00, the shorter hand would point to the "2."
6	A	On a clock, the shorter hand points toward the hour and the longer hand points toward the minutes. For example, if it was 2:30, the longer hand would point to the "6," which represents the 30th minutes.
7	C	The hour hand (the shorter hand) is pointed at the 10th hour, and the minute hand (the longer hand) is at 2 minutes. The clock shows 10:02. Eight minutes after 10:02 would be 10:10.
8	A	The hour hand (the shorter hand) is past the 12th hour but not yet at the 1st hour, and the minute hand (the longer hand) is at 39 minutes. The clock shows 12:39. Twenty minutes after 12:39 would be 12:59.
9	B	The hour hand (the shorter hand) is past the 12th hour but not yet at the 1st hour, and the minute hand (the longer hand) is at 39 minutes. The clock shows 12:39. Ten minutes before 12:39 would be 12:29.

© Lumos Information Services 2018 | LumosLearning.com

Question No.	Answer	Detailed Explanation
10	B.	Lucy's time:12:58 - 12:09 = 51 minutes. David's time: 1:03 - 12:15 = 48 minutes. David has the shorter time.
11	C	You can solve this problem by converting the hours to minutes and then subtracting the two times. 5 hours and 25 minutes is equivalent to (5 x 60) + 25 = 325 minutes. You multiply 5 hours by 60 because there are 60 minutes in an hour. 3 hours and 41 minutes is equivalent to (3 x 60) + 41 = 221 minutes. 325 - 221 = 104. Now convert 104 minutes back into hours and minutes by dividing by 60 and the answer is 1 hour and 44 minutes.
12	C	To calculate how many hours of sleep Rachel will receive, add the amount of time she sleeps to the time she goes to bed. 9 hours after 9:30 PM would be 6:30 AM.
13	B	To calculate when the show began, subtract the length of the show from the ending time. Counting back 45 minutes from 12:20 PM, you would arrive at 11:35 AM. The PM changes to AM, since you are now before noon.
14	C	To calculate when the students have to be finished with their test, add the amount of time given for the test to the start time. 40 minutes after 10:22 AM would be 11:02 AM.
15	A	To calculate when the pizza will be done, add the cooking time to the time Mr. Adams began cooking. 25 minutes after 5:43 PM would be 6:08 PM.

Lesson 2: Elapsed Time

Question No.	Answer	Detailed Explanation
1	A	Counting back from 10:02 to 10:00 is 2 minutes. Then, counting back from 10:00 back to 9:12 is an additional 48 minutes, making the total elapsed time 50 minutes.
2	C	Subtract the beginning time from the ending time; 7:35 back to 7:11 is 24 minutes.
3	D	Subtract the time Tanya began walking from the time she arrived home; 3:56 back to 3:31 is 25 minutes.

© Lumos Information Services 2018 | LumosLearning.com

Question No.	Answer	Detailed Explanation
4	B	From 4:00 to 5:00 is one hour of elapsed time. From 5:00 until 5:27 is an additional 27 minutes, for a total elapsed time of 1 hour and 27 minutes.
5	C	From 6:12 to 7:12 is 1 hour of elapsed time. From 7:12 to 7:15 is an additional 3 minutes, for a total elapsed time of 1 hour and 3 minutes.
6	D	From 7:06 to 8:06 is one hour of elapsed time. From 8:06 to 8:13 is an additional 7 minutes, for a total elapsed time of 1 hour and 7 minutes.
7	B	From 3:11 to 4:11 is one hour of elapsed time. From 4:11 to 4:37 is an additional 26 minutes, for a total elapsed time of 1 hour and 26 minutes.
8	A	Counting back from 9:03 to 9:00 is 3 minutes. Then, counting back from 9:00 to 8:19 is an additional 41 minutes, for a total elapsed time of 44 minutes.
9	A	Subtract the beginning time from the ending time; 4:32 back to 4:17 is 15 minutes.
10	D	Counting back from 4:11 to 4:00 is 11 minutes. Then, counting back from 4:00 to 3:28 is an additional 32 minutes, for a total elapsed time of 43 minutes.
11	B	Subtract the present time from the time he has to leave; 1:30 - 1:11 = 19 minutes
12	A	From 10:45 to 11:00 is 15 minutes of elapsed time. From 11:00 to 11:25 is an additional 25 minutes, for a total elapsed time of 40 minutes.
13	A	Counting back from noon (12:00) to 11:30 is 30 minutes. Then, counting back from 11:30 to 11:29 is an additional minute, for a total elapsed time of 31 minutes.
14	A	Clock A shows 12:01 and Clock B shows 1:03. The elapsed time from 12:01 to 1:03 is 1 hour and 2 minutes.
15	D	Clock A shows 7:15 and Clock B shows 7:47. The elapsed time from 7:15 to 7:47 is 32 minutes.

© Lumos Information Services 2018 | LumosLearning.com

Lesson 3: Liquid Volume & Mass

Question No.	Answer	Detailed Explanation
1	D	The word "pound" after the number 40 indicates that 40 is a measurement of weight.
2	B	Kilograms are used to measure the mass of large, solid objects such as a table.
3	A	The amount of a liquid is also called its volume. Option C and D are not used to measure volume. Option B would be too large to measure water in a small bowl. Option A is the only logical choice.
4	B	Options A and C would not be used to measure weight. Option D would be too small to measure the weight of a table. Option B is the most logical choice.
5	A	Capacity is another word for volume. Options B, C, and D would not be used to measure volume. Option A is the most logical choice.
6	C	Option A is a measure for volume. Option B is a measure of distance or length. Option D is a measure of temperature. Option C is the most logical choice.
7	A	A gallon is a unit used to measure the volume of large amounts of liquid, whereas the other units measure smaller amounts of liquids.
8	C	Options A and B would be too light to be the weight of a typical 8-year-old, While Option D would be too heavy. Option C is the most logical choice.
9	A	Volume is a measurement associated with 3-dimensional figures. As a result, it is represented in cubic units. Volume is a measure of how many identical cubes (or parts of identical cubes) could fit within a solid figure.
10	B	Options C and D are not used to measure weight. Option A would be used to measure the weight of very large objects. Option B is the most logical choice, for the large amount of water that takes up a swimming pool.
11	C	Option D is not used to measure volume. Options A and B are both used to measure small amounts of liquids. Option C is the most logical choice for the large amount of water that takes up a swimming pool.

© Lumos Information Services 2018 | LumosLearning.com

Question No.	Answer	Detailed Explanation
12	D	Options A, B, and C are all used to measure shorter lengths. Option D is the most logical choice.
13	A	Option C is not used to measure mass or volume. A cupcake recipe would have a very small amount of salt. Options B and D are both used to measure large quantities. Option A is the most logical choice for the small amount of salt that would be in Cupcakes.
14	A	Options B, C, and D are all used to measure shorter lengths and distances. Option A is used to measure large distances.
15	C	A millimeter is a very small unit of length (approximately the thickness of a fingernail).

Lesson 4: Graphs

Question No.	Answer	Detailed Explanation
1	B	The chart shows 3 sets of 5 tallies for Mrs. B's class in the "yes" column. Multiplying 3 x 5 the tallies represent 15 kids.
2	C	The number 14 in the "yes" column for Mrs. A's class represents 14 votes.
3	D	The "Total" row displays the overall number of votes. There is a total of 41 votes represented in the "No" column.
4	C	First, add the totals number of students who chose Science and Math. 3 + 4 = 7. The chart states that each object stands for 2 votes. Multiply the Science and Math total by 2. 7 x 2 = 14.
5	C	The tallest bar indicates the food that was chosen most often. That would be considered the "favorite."
6	D	The foods with the shortest bars represent the foods that were least liked by the kids. Both salads and wraps had the least amount of votes.
7	C	Locate "pasta" on the bottom of the graph. The bar for pasta reaches up to the 40 line.

© Lumos Information Services 2018 | LumosLearning.com

Question No.	Answer	Detailed Explanation
8	B	First find the values for both fries and pizza by locating them on the x-axis and then moving over to the y-axis to see their value. Subtract the number of kids who chose pizza from the number of kids who chose fries. 50 - 40 = 10.
9	B	The title of the graph is located above the graph.
10	A	Option B is false because the graph makes no mention of amount of rain. Option C is false because the title of the graph states "rainy days" and not temperature. Option D is false because the graph only shows 5 months which is not equivalent to a year. Option A is the only choice that is true.
11	B	According to the graph, January had 2 rainy days, February had 8 rainy days, March had 4 rainy days, and April had 6 rainy days. February had the most rainy days.
12	B	Add up all the tallies to obtain the total. 6 + 4 + 8 + 4 + 3 = 25.
13	C	Subtract the number of students who chose hockey from the number of students who chose soccer. 6 - 4 = 2.
14	B	There are 8 tally marks in the baseball section. This represents the 8 students who voted for baseball as their favorite sport.
15	D	Hockey and tennis both have 4 tallies on the chart.

Lesson 5: Measuring Length

Question No.	Answer	Detailed Explanation
1	C	Options A, B, and D are all customary units. Option C is the only metric unit.
2	B	A meter is just slightly longer than a yard. That is why a meterstick and a yardstick are almost the same length. 1 yard = 36 inches. 1 meter ≈ 39 inches.
3	C	Decimeters and centimeters are both too small to measure a table. Kilometers are used to measure long distances or lengths. Option C is the most appropriate.
4	C	Kilometers and meters are used to measure longer lengths. Grams are used to measure mass. Option C is the most appropriate.

© Lumos Information Services 2018 | LumosLearning.com

Question No.	Answer	Detailed Explanation
5	A	Feet, yards, and miles are all used to measure longer lengths. The length of a pencil would be measured in inches.
6	A	Options B, C, and D are all too long to be the measure of a football. Option A is the most appropriate.
7	B	Option A and C are both too small for a football field. Option D is inappropriate because a football field is measurable. Option B is the most appropriate answer.
8	D	The width of a finger is small. Options A and B would both be too large. Option C is too small and would be more appropriate for the thickness of a fingernail. Option D is the most appropriate.
9	B	Hours are used to measure time. Yards are too small to measure a distance between two cities. Square inches are a measure of area, not distance. Option B is the most appropriate.
10	C	There are 2.54 centimeters in an inch. To convert centimeters to inches, divide 25 ÷ 2.54 = 9.84, rounding 9.84 to 10.
11	B	This ruler measures in inches. The end of the object stops at the number 8 so this object must be 8 inches.
12	C	This ruler measures in inches. The end of the object stops at the number 3 so this object must be 3 inches.
13	A	This ruler measures in inches. The end of the object stops at the number 6 so this object must be 6 inches.
14	B	Option A is false because inches are a smaller measurement unit than miles. Option C is false because a foot is longer than an inch. Option D is false because a mile is larger than a foot. Option B is the only statement that is true.
15	C	There are exactly 12 inches in one foot.

Lesson 6: Area

Question No.	Answer	Detailed Explanation
1	C	Area is a 2-dimensional attribute, so it must be represented in square units. Area is a measure of how many identical squares (or parts of identical squares) would be needed to cover a figure.

© Lumos Information Services 2018 | LumosLearning.com

Question No.	Answer	Detailed Explanation
2	D	A postage stamp is a rectangle measuring about 1 inch on each side. Therefore, the area of a postage stamp is about 1 square inch. The other objects are all too large to measure 1 square inch as this is a very small measurement.
3	D	The graphing paper has the largest amount of square units if each box represents 1 unit. A toy block is a 3-dimensional cube-shape and cannot be measured in square units.
4	C	If each box is a square unit, count the number of shaded boxes to get the area of the shaded region. There are 11 shaded boxes so the area is equal to 11 square units.
5	D	If each box is a square unit, count the number of boxes to get the area. There are 16 boxes so the area is equal to 16 square units.
6	C	If each box is a square unit, count the number of boxes to get the area. There are 24 boxes so the area is equal to 24 square units.
7	B	If each box is a square unit, count the number of shaded boxes to get the area of the shaded region. There are 10 shaded boxes so the area is equal to 10 square units.
8	B	If each box is a square unit, count the number of shaded boxes to get the area of the shaded region. There are 12 shaded boxes so the area is equal to 12 square units.
9	D	If each box is a square unit, count the number of shaded boxes to get the area of the shaded region. There are 9 shaded boxes so the area is equal to 9 square units.
10	C	If each box is a square unit, count the number of shaded boxes to get the area of the shaded region. There are 12 shaded boxes so the area is equal to 12 square units.
11	A	If each box is a square unit, count the number of shaded boxes to get the area of the shaded region. There are 10 shaded boxes so the area is equal to 10 square units.
12	B	If each box is a square unit, count the number of boxes to get the area. There are 36 boxes so the area is equal to 36 square units.
13	B	Figure B contains 24 square units while figure A only contains 12 square units. The area of figure B is larger than that of figure A.

Question No.	Answer	Detailed Explanation

© Lumos Information Services 2018 | LumosLearning.com

14	B	Figure A has 10 square units. Figure C also has 10 square units. Figure B only has 8 square units.
15	D	If each box is a square unit, count the number of boxes to get the area. There are 16 boxes so the area is equal to 16 square units.

Lesson 7: Relating Area to Addition & Multiplication

Question No.	Answer	Detailed Explanation
1	A	Area is calculated by multiplying length by width: 3 feet x 29 feet = 87 square feet. (Note: 3 x 29 = 29 + 29 + 29 = 87)
2	D	Area is calculated by multiplying length by width: 12 yards x 15 yards = 180 square yards. [Note: 15 x 12 = (15 x 10) + (15 x 2) = 150 + 30 = 180]
3	C	Area of a rectangle is calculated by multiplying length by width. To find the area of this rectangle, multiply 63 x 33.
4	B	Area of a rectangle is calculated by multiplying length by width: 5 meters x 10 meters = 50 square meters.
5	A	Area of a rectangle is calculated by multiplying length by width: 16 yards x 11 yards = 176 square yards. [Note: 16 x 11 = (16 x 10) + (16 x 1) = 160 + 16 = 176]
6	B	Area of a rectangle is calculated by multiplying length by width: 3 inches x (2 + 3) inches = 3 inches x 5 inches = 15 square inches.
7	D	Area of a rectangle is calculated by multiplying length by width: 7 feet x (2 + 1) feet = 7 feet x 3 feet = 21 square feet.
8	A	Area of a rectangle is calculated by multiplying length by width: 12 meters x (4 + 3) meters = 12 meters x 7 meters = 84 square meters.
9	D	Area of a rectangle is calculated by multiplying length by width: 5 inches x (5 + 2) inches = 5 inches x 7 inches = 35 square inches.
10	B	Area of a rectangle is calculated by multiplying length by width: 13 yards x (7 x 7) yards = 13 yards x 14 yards = 182 square yards.

© Lumos Information Services 2018 | LumosLearning.com

Question No.	Answer	Detailed Explanation
11	C	The park is rectangular. The area of a rectangle is calculated by multiplying length by width: 50 feet x 20 feet = 1,000 square feet.
12	A	Area of a rectangle is calculated by multiplying length by width: 7 yards x 6 yards = 42 square yards.
13	C	Area of a rectangle is calculated by multiplying length by width: 30 feet x 20 feet = 600 square feet.
14	C	The area of one wall is calculated by multiplying length by width: 20 feet x 10 feet = 200 square feet. There are four identical walls, so multiply 4 by 200 to calculate the total area: 4 x 200 square feet = 800 square feet.
15	C	Area is calculated by multiplying length by width: 2 meters x 4 meters = 8 square meters.
		Lesson 8: Perimeter
1	C	The perimeter of a shape is the distance around the shape.
2	B	A plane figure is a two-dimensional (flat) shape. Area and perimeter are both associated with these types of objects. Volume and weight apply to 3-dimensional figures.
3	B	The perimeter of a plane figure is calculated by adding the lengths of the four sides. Area is measured by totaling the square units, or multiplying the length by the width. Diameter and circumference refer to measurements of a circle.
4	C	The formula a + b + a + b= can represent the four sides of a rectangle being added together. The other choices. Each set of two sides is always equal. The formula a x b refers to area, not perimeter.
5	A	To find the perimeter of a rectangle, total the lengths of its four sides. 4 + 1 + 4 + 1 = 10 units

© Lumos Information Services 2018 | LumosLearning.com

Question No.	Answer	Detailed Explanation
6	B	To find the perimeter of a rectangle, total the lengths of its four sides. 1 + 2 + 1 + 2 = 6 units
7	A	To find the perimeter of the shaded area, count the edges that surround the outside of the shaded area. There are 16 sides around the shaded area so the perimeter is 16.
8	C	A rhombus has 4 equal sides and the perimeter is equal to the sum of those 4 sides. To calculate the length of each side, divide the perimeter by 4. 20 ÷ 4 = 5.
9	B	Because a rhombus has four equal sides, to find its perimeter multiply the length of one of its sides by four. 3 x 4 = 12 centimeters
10	C	A square has 4 equal sides and the perimeter is equal to the sum of those 4 sides. To calculate the length of each side, divide the perimeter by 4. 80 ÷ 4 = 20.
11	A	To find the perimeter of the figure, count the edges that surround the outside of the figure. There are 20 sides around the figure so the perimeter is 20.
12	B	The yard is rectangular. To find its perimeter, total the lengths of its four sides. 30 + 15 + 30 + 15 = 90 feet
13	A	To find the perimeter of the figure, count the edges that surround the outside of the figure. There are 20 sides around the figure so the perimeter is 20.
14	C	The field is rectangular. To find its perimeter, add the lengths of its four sides. 23 + 32 + 23 + 32 = 110 yards
15	A	To find the perimeter of a rectangle, total the lengths of its four sides. 15 + 12 + 15 + 12 = 54 feet
16	D	To find the perimeter of a rectangle, total the lengths of its four sides. 16 + 11 + 16 + 11 = 54 feet

© Lumos Information Services 2018 | LumosLearning.com

Question No.	Answer	Detailed Explanation
17	D	To find the perimeter of the shaded area, count the edges that surround the outside of the shaded area. There are 16 sides around the shaded area so the perimeter is 16.
18	D	To find the perimeter of a rectangle, total the lengths of its four sides. 3 + 29 + 3 + 29 = 64 feet
19	C	The park is rectangular. To find its perimeter, total the lengths of its four sides. 20 + 50 + 20 + 50 = 140 feet
20	B	To find the perimeter of a rectangle, total the lengths of its four sides. 33 + 53 + 33 + 53 = 172 inches
21	C	To find the perimeter of a rectangle, total the lengths of its four sides. 10 + 5 + 10 + 5 = 30 meters
22	A	To solve for the missing side, plug in what you know into the formula for perimeter: 2W + 2L = P (2 x 3) + 2x = 16 To solve for x: 2x = 16 - 6 = 10. Divide 10 by 2 to find one side; x = 5 feet.
23	A	To solve for the missing side, plug in what you know into the formula for perimeter: 2W + 2L = P (2 x 7) + 2x = 20 To solve for x: 2x = 20 - 14 = 6 Divide 6 by 2 to find one side; x = 3 feet.
24	D	If there are 4 sides to the rectangular park, the measurements of each side would be 20, 50, 20, 50. The question asks for the least amount of fencing needed for 3 sides. They should choose to fence the two 20-foot sides and one of the 50-foot sides. The amount of fencing material needed would be 20 + 20 + 50 = 90 feet.
25	C	To solve for the missing side, plug in what you know into the formula for perimeter: 2W + 2L = P. (2 x 12) + 2x = 38, now solve for x. To solve for x: x = (38 - 24) ÷ 2, x = 7.

© Lumos Information Services 2018 | LumosLearning.com

Question No.	Answer	Detailed Explanation
26	B	To solve for the missing side, plug in what you know into the formula for perimeter: 2W + 2L = P (2 x 5) + 2x = 24 To solve for x : 2x = 24 - 10 = 14 Divide 14 by 2 to find one side; x = 7 inches.
27	B	To solve for the missing side, plug in what you know into the formula for perimeter: 2W + 2L = P (2 x 13) + 2x = 54 To solve for x: 2x = 54 - 26 = 28 Divide 28 by 2 to find one side; x = 14 yards.
28	A	Lisa used a 13" x 9" rectangular cake pan. 9 + 13 + 9 + 13 = 44. The other choices do not total 44 when the sides are added. A round cake pan is a circle and does not have 4 sides, so choice D is not logical.
29	A	7+15+7+15= 44. None of the other figures have a perimeter of 44 when the 4 sides are added together.
30	C	20+8+20+8= 56 None of the other figures have a perimeter of 56 when the 4 sides are added together.

© Lumos Information Services 2018 | LumosLearning.com

Name ________________________ Date ______________

Chapter 5: Geometry

Lesson 1: 2-Dimensional Shapes

You can scan the QR code given below or use the url to access additional EdSearch resources including videos and mobile apps related to *2-Dimensional Shapes*.

Categories	About 7 results (0.008 seconds)
Apps (3)	Categorize quadrilaterals
Questions (2)	**Resource:** Khan Academy
Khan Academy (1)	**Standard:** 3.G.A.1
Videos (1)	**Grade:** 3
Popular Searches	**Subject:** Math
Recent Searches	Topic Standard

edSearch **_2-Dimensional Shapes_**

URL	QR Code
http://www.lumoslearning.com/a/3ga1	

Name ______________________ Date ______________

1. **Fill in the blank with the correct term.**
   Closed, plane figures that have straight sides are called ________ .

   Ⓐ parallelograms
   Ⓑ line segments
   Ⓒ polygons
   Ⓓ squares

2. **Which of the following shapes is not a polygon?**

   Ⓐ Square
   Ⓑ Hexagon
   Ⓒ Circle
   Ⓓ Pentagon

3. **Complete this statement.**
   A rectangle must have ___________ .

   Ⓐ four right angles
   Ⓑ four straight angles
   Ⓒ four obtuse angles
   Ⓓ four acute angles

4. **How many sides does a trapezoid have?**

   Ⓐ 4
   Ⓑ 8
   Ⓒ 6
   Ⓓ 10

5. **Complete the following statement.**
   A square is always a ________ .

   Ⓐ rhombus
   Ⓑ parallelogram
   Ⓒ rectangle
   Ⓓ All of the above

© Lumos Information Services 2018 | LumosLearning.com

Name ______________________________ Date ______________

6. Which of these statements is true?

   Ⓐ A square and a triangle have the same number of angles.
   Ⓑ A triangle has more angles than a square.
   Ⓒ A square has more angles than a triangle.
   Ⓓ A square and a triangle each have no angles.

7. Which of these statements is true?

   Ⓐ A rectangle has more sides than a trapezoid.
   Ⓑ A parallelogram and a trapezoid have the same number of sides.
   Ⓒ A triangle has more sides than a trapezoid.
   Ⓓ A triangle has more sides than a square.

8. Complete this statement.
   A trapezoid must have ____________.

   Ⓐ two acute angles
   Ⓑ two right angles
   Ⓒ one pair of parallel sides
   Ⓓ two pairs of parallel sides

9. Complete the following statement.
   Squares, rectangles, rhombi and trapezoids are all _________.

   Ⓐ triangles
   Ⓑ quadrilaterals
   Ⓒ angles
   Ⓓ round

10. Which of these shapes is a quadrilateral?

    Ⓐ circle
    Ⓑ triangle
    Ⓒ rectangle
    Ⓓ pentagon

11. Which of these shapes is NOT a quadrilateral?

    Ⓐ square
    Ⓑ trapezoid
    Ⓒ rectangle
    Ⓓ triangle

© Lumos Information Services 2018 | LumosLearning.com

Name ______________________ Date ______________

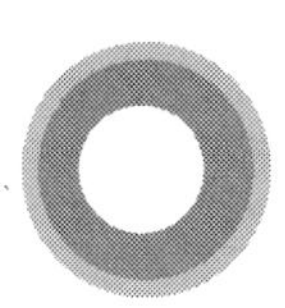

12. Name the figure shown below.

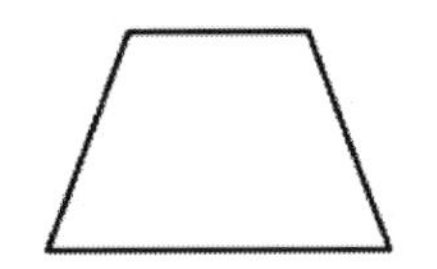

Ⓐ Trapezoid
Ⓑ Square
Ⓒ Pentagon
Ⓓ Rhombus

13. Name the object shown below.

Ⓐ Rectangle
Ⓑ Parallelogram
Ⓒ Trapezoid
Ⓓ Rhombus

14. The figure shown below is a __________ .

Ⓐ parallelogram
Ⓑ rectangle
Ⓒ quadrilateral
Ⓓ All of the above

15. The figure below is a __________ .

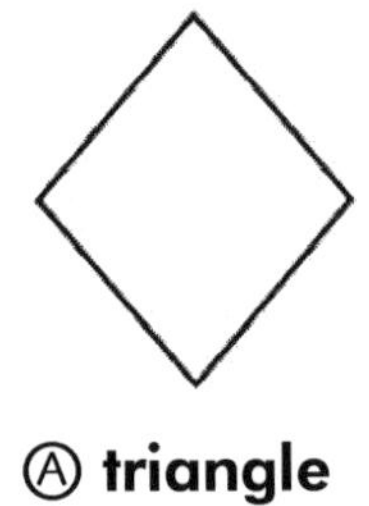

Ⓐ triangle
Ⓑ square
Ⓒ rhombus
Ⓓ trapezoid

© Lumos Information Services 2018 | LumosLearning.com

Chapter 5

Lesson 2: Shape Partitions

You can scan the QR code given below or use the url to access additional EdSearch resources including videos and mobile apps related to *Shape Partitions*.

© Lumos Information Services 2018 | LumosLearning.com

Name ______________________ Date ______________

1. What is the dotted line that divides a shape into two equal parts called?

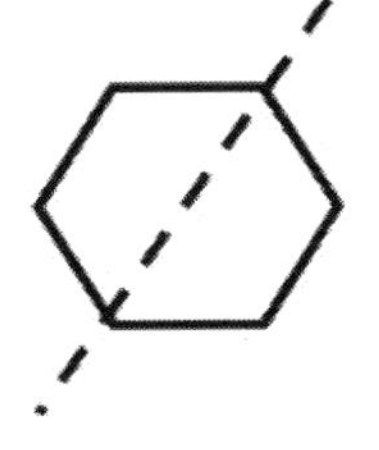

Ⓐ a middle line
Ⓑ a line of symmetry
Ⓒ a line of congruency
Ⓓ a divider

2. A square has how many lines of symmetry?

Ⓐ 8
Ⓑ 4
Ⓒ 1
Ⓓ 2

3. Which of the following has NO lines of symmetry?

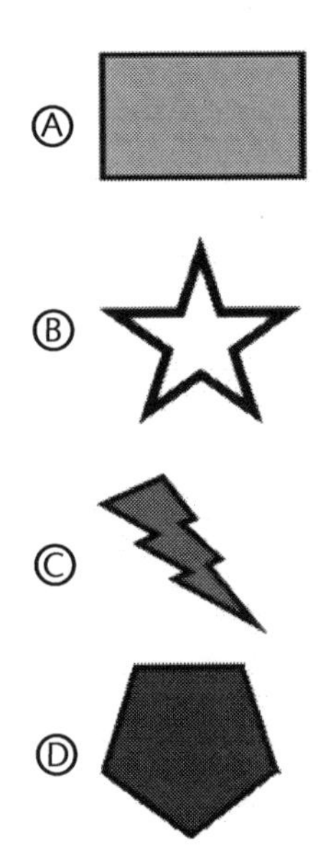

© Lumos Information Services 2018 | LumosLearning.com

4. **Which of the following objects has more than one line of symmetry?**

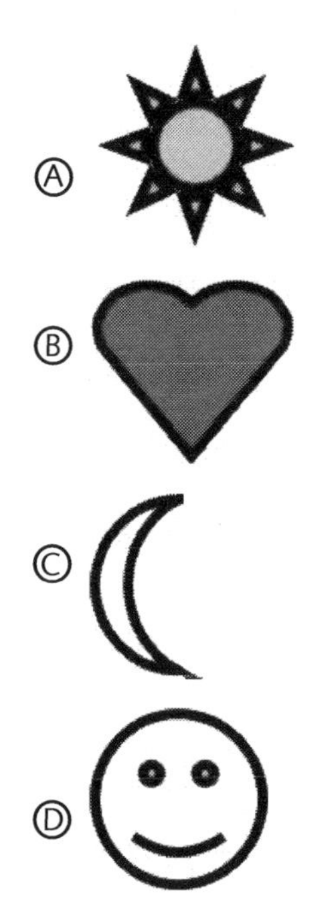

Ⓐ

Ⓑ

Ⓒ

Ⓓ

5. **What fraction of this triangle is shaded?**

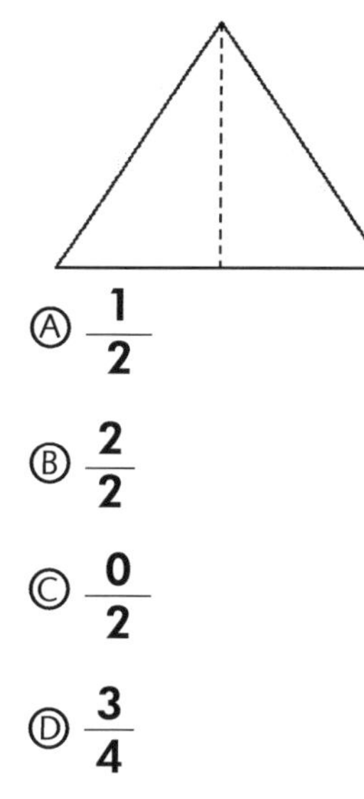

Ⓐ $\frac{1}{2}$

Ⓑ $\frac{2}{2}$

Ⓒ $\frac{0}{2}$

Ⓓ $\frac{3}{4}$

© Lumos Information Services 2018 | LumosLearning.com

Name ______________________ Date ______________

6. **What fraction of this triangle is shaded?**

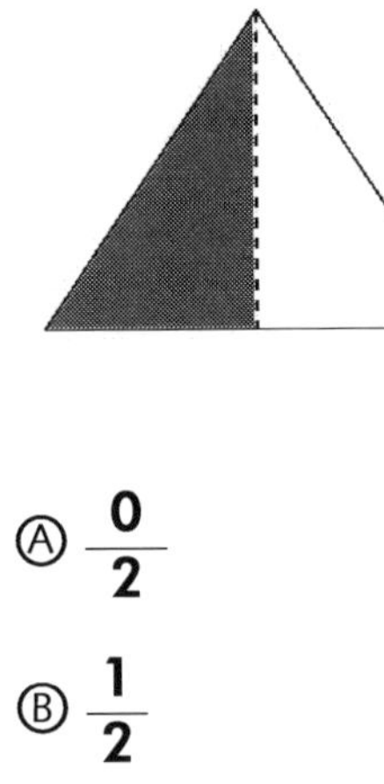

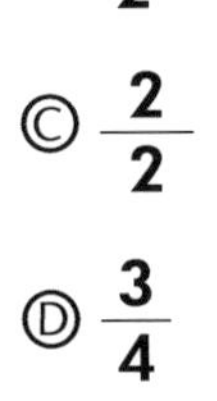

Ⓐ $\frac{0}{2}$

Ⓑ $\frac{1}{2}$

Ⓒ $\frac{2}{2}$

Ⓓ $\frac{3}{4}$

7. **What fraction of this triangle is shaded?**

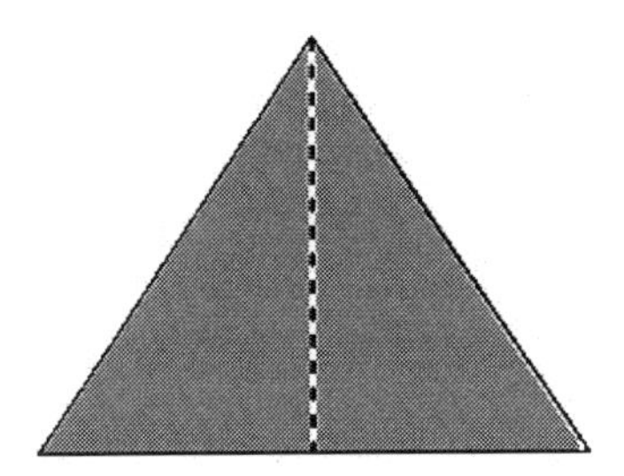

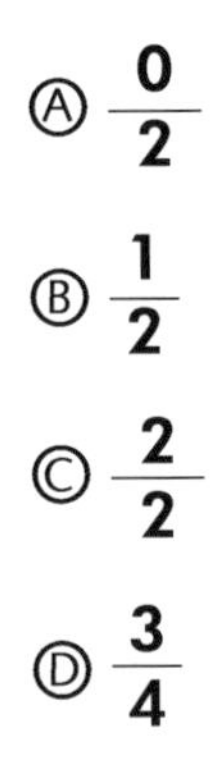

Ⓐ $\frac{0}{2}$

Ⓑ $\frac{1}{2}$

Ⓒ $\frac{2}{2}$

Ⓓ $\frac{3}{4}$

© Lumos Information Services 2018 | LumosLearning.com

8. **What fraction of this square is shaded?**

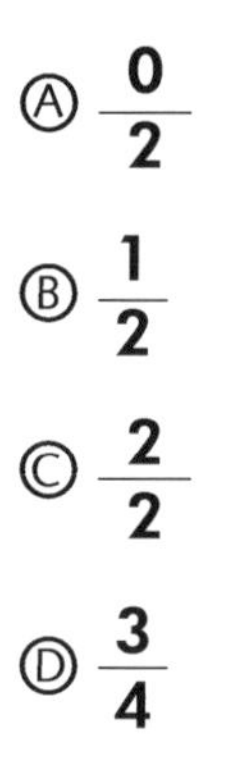

9. **What fraction of this square is shaded?**

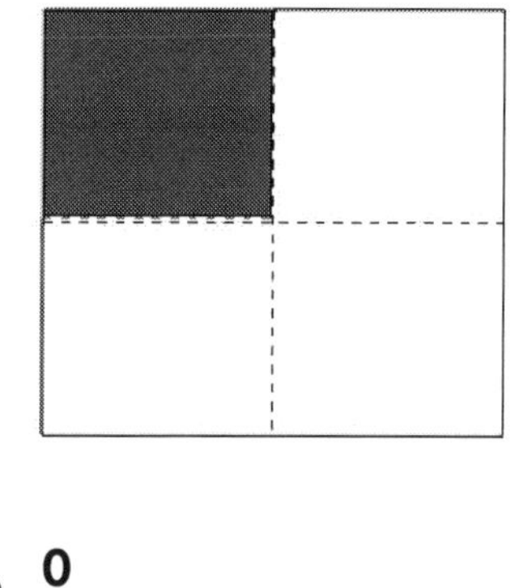

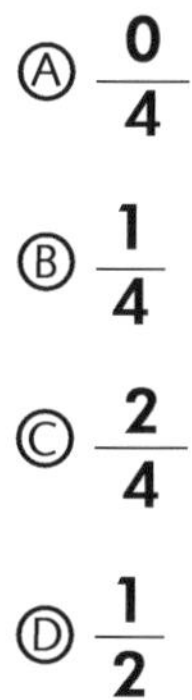

© Lumos Information Services 2018 | LumosLearning.com

Name ______________________ Date ______________

10. What fraction of this square is shaded?

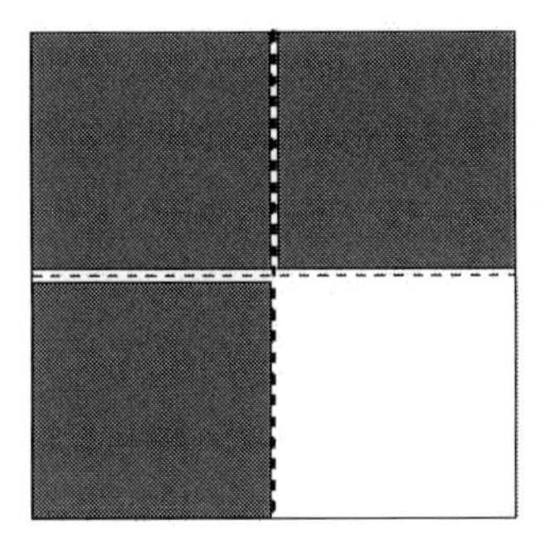

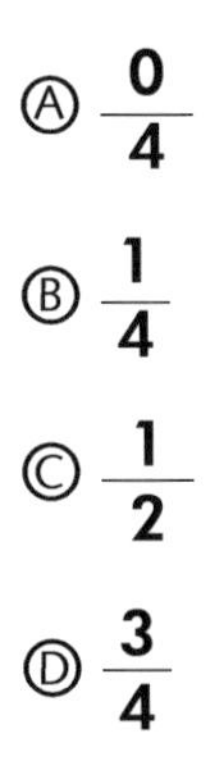

Ⓐ $\frac{0}{4}$

Ⓑ $\frac{1}{4}$

Ⓒ $\frac{1}{2}$

Ⓓ $\frac{3}{4}$

11. What fraction of this rectangle is shaded?

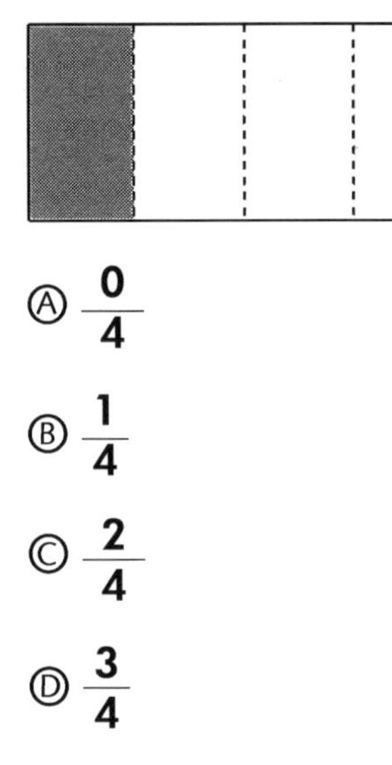

Ⓐ $\frac{0}{4}$

Ⓑ $\frac{1}{4}$

Ⓒ $\frac{2}{4}$

Ⓓ $\frac{3}{4}$

12. What fraction of this circle is shaded?

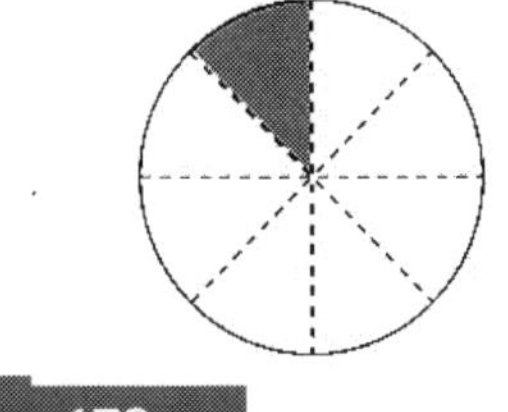

© Lumos Information Services 2018 | LumosLearning.com

Ⓐ $\frac{1}{8}$

Ⓑ $\frac{1}{4}$

Ⓒ $\frac{1}{2}$

Ⓓ $\frac{3}{4}$

13. What fraction of this circle is shaded?

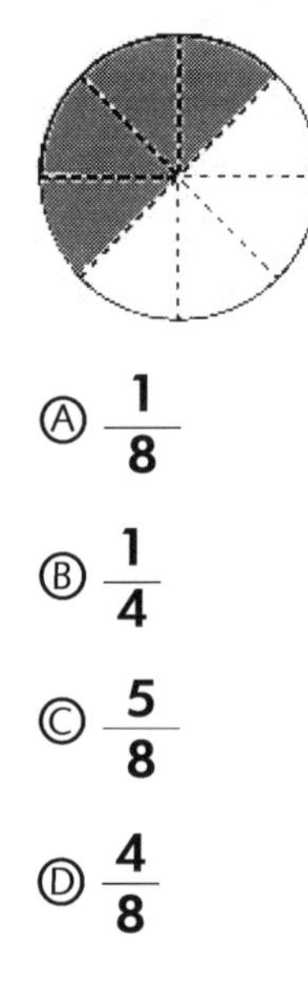

Ⓐ $\frac{1}{8}$

Ⓑ $\frac{1}{4}$

Ⓒ $\frac{5}{8}$

Ⓓ $\frac{4}{8}$

14. The area of the entire rectangle shown below is 48 square feet. What is the area of the shaded portion?

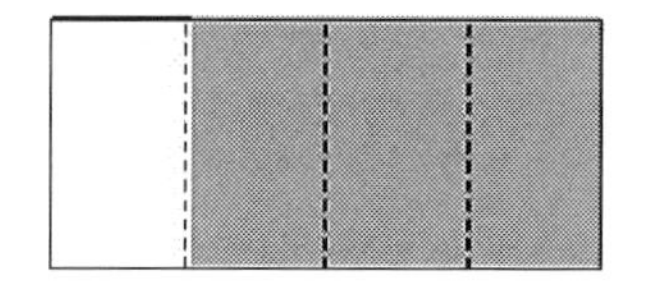

Ⓐ 36 square feet
Ⓑ 48 square feet
Ⓒ 144 square feet
Ⓓ 12 square feet

© Lumos Information Services 2018 | LumosLearning.com

15. If the area of the entire rectangle below is 36 square feet. What is the area of the shaded portion?

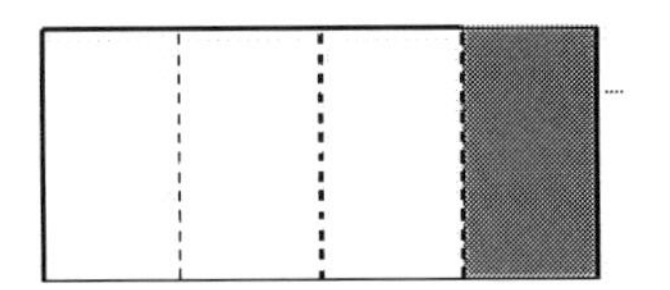

Ⓐ 8 square feet
Ⓑ 9 square feet
Ⓒ 144 square feet
Ⓓ 12 square feet

End of Geometry

© Lumos Information Services 2018 | LumosLearning.com

Chapter 5: Geometry

Answer Key
&
Detailed Explanations

© Lumos Information Services 2018 | LumosLearning.com

Name ______________________ Date ______________

Lesson 1: 2-Dimensional Shapes

Question No.	Answer	Detailed Explanation
1	C	By definition, a polygon is a plane (flat), closed figure with only straight sides.
2	C	A polygon must have only straight sides. A circle is the only option that does not fit this criteria. Since it is curved.
3	A	A rectangle is a quadrilateral (4-sided polygon) with 4 right angles.
4	A	A trapezoid is a quadrilateral which means it has 4 sides.
5	D	A square is a rhombus parallelogram, and rectangle because all of these figures are four-sided and contain two sets of parallel sides.
6	C.	Option C is true because a square has 4 angles and a triangle has 3 angles.
7	B	Option B is true because a parallelogram and a trapezoid both have 4 sides.
8	C	A trapezoid is a quadrilateral with one pair of parallel sides.
9	B	A quadrilateral is a figure with four straight sides and four angles. Squares, rectangles, rhombi, and trapezoids all have 4 sides.
10	C	A quadrilateral is a figure with four straight sides and four angles. A rectangle fits this description, where as a triangle has 3 sides, a circle is round, and a pentagon has 5 sides.
11	D	A quadrilateral is a figure with four straight sides and four angles. A triangle is the only choice that does not fit this description. Since it has only 3 sides.
12.	A	A trapezoid is a quadrilateral that contains only one pair of parallel sides.
13	B	A parallelogram is a quadrilateral with two pairs of opposite parallel sides. The figure is not a rectangle because it does not have right angles. It is not a rhombus because the sides are not all equal in length. It is not a trapezoid because a trapezoid only has one set of parallel sides.
14	D	The shape is a parallelogram because it has two pairs of parallel sides. It is a quadrilateral because it has 4 sides. It is a rectangle because it is a parallelogram with all right angles.
15	C	A rhombus is a quadrilateral with 4 equal sides. This figure is not a square because it does not have right angles. Triangles are three-sided, while trapezoids do not have four equal sides.

© Lumos Information Services 2018 | LumosLearning.com

Lesson 2: Shape Partitions

Question No.	Answer	Detailed Explanation
1	B	A line of symmetry is an imaginary line that divides an object into two mirror images.
2	B	A line of symmetry is an imaginary line that divides an object into two mirror images. A square can be divided across the length, across the width, down diagonally from left to right, and down diagonally from right to left.
3	C	A line of symmetry is an imaginary line that divides an object into two mirror images. Option C cannot be divided in such a way.
4	A	A line of symmetry is an imaginary line that divides an object into two mirror images. The object in Option A can have a line of symmetry at multiple points, for example across the length, across the width, and diagonally.
5	C	No parts of the triangle are shaded yet the shape is divided into two parts. To form a fraction, the numerator is the part and the denominator is the whole. Since no parts are shaded, the numerator would be 0.
6	B	One part of the triangle is shaded yet the shape is divided into two parts. To form a fraction, the numerator is the part and the denominator is the whole.
7	C	Two parts of the triangle are shaded and the shape is divided into two parts. To form a fraction, the numerator is the part and the denominator is the whole.
8	B	One part of the square is shaded and the shape is divided into two parts. To form a fraction, the numerator is the part and the denominator is the whole.
9	B	One part of the square is shaded and the shape is divided into four parts. To form a fraction, the numerator is the part and the denominator is the whole.
10	D	Three parts of the square are shaded and the shape is divided into four parts. To form a fraction, the numerator is the part and the denominator is the whole.
11	B	One part of the rectangle is shaded yet the shape is divided into four parts. To form a fraction, the numerator is the parts and the denominator is the whole.

© Lumos Information Services 2018 | LumosLearning.com

Name ______________________ Date ______________

Question No.	Answer	Detailed Explanation
12	A	One part of the circle is shaded yet the shape is divided into eight parts. To form a fraction, the numerator is the parts and the denominator is the whole.
13	D	Four parts of the circle are shaded yet the shape is divided into eight parts. To form a fraction, the numerator is the parts and the denominator is the whole.
14	A	The shape is divided into four equal parts. This means that each part has the same area. If the total area is known, divide this area by 4 to calculate the area of each part. $48 \div 4 = 12$. Since one part is left unshaded, subtract 12 from the total area of 48 to find that the shaded portion represents 36 square feet.
15	B	The shape is divided into four equal parts. This means that each part has the same area. If the total area is known, divide this area by 4 to calculate the area of each part. $36 \div 4 = 9$. Each piece has an area of 9 square feet. The problem asks for the area of the shaded portion. There is only one shaded section, so the area equals 9 square feet.

© Lumos Information Services 2018 | LumosLearning.com

Additional Information

PARCC FAQs

What Will PARCC Math Assessments Look Like?

In many ways, the PARCC assessments will be unlike anything many students have ever seen. The tests will be conducted online, requiring students complete tasks to assess a deeper understanding of the CCSS. The students will be assessed once 75% of the year has been completed in one Summative based assessment and the Summative Assessment will be broken into:
Unit 1, Unit 2, Unit 3, and Unit 4 for Grades 3 to 5 and Unit 1, Unit 2 and Unit 3 for Grades 6 to 8.

For Math, PARCC differentiates three different types of questions:

Type I – Tasks assessing concepts, skills, procedures

- Balance of conceptual understanding, fluency, and application
- Can involve any or all mathematical practice standards
- Machine scorable including innovative, computer-based formats

Type II - Tasks assessing expressing mathematical reasoning

- Each task calls for written arguments/justifications, critique of reasoning or precision in mathematical statements.
- Can involve other mathematical practice standards
- May include a mix of machine-scored and hand-scored responses

Type III - Tasks assessing modeling/applications

- Each task calls for modeling/application in a real-world context or scenario
- Can involve other mathematical practice standards
- May include a mix of machine-scored and hand-scored responses

What will PARCC Look Like?

In many ways, the PARCC tests will be unlike anything many students have ever seen. The tests will be conducted online, requiring students complete tasks to assess a deeper understanding of the Common Core State Standards. The students will take the Test at the end of the year.

The time for each Math unit is described below:

Estimated Time on Task in Minutes				
Grade	**Unit 1**	**Unit 2**	**Unit 3**	**Unit 4**
3	60	60	60	60
4	60	60	60	60
5	60	60	60	60
6	80	80	80	NA
7	80	80	80	NA
8	80	80	80	NA

How is this Lumos tedBook aligned to PARCC Guidelines?

Although the PARCC assessments will be conducted online, the practice tests here have been created to accurately reflect the depth and rigor of PARCC tasks in a pencil and paper format. Students will still be exposed to the TECR technology style questions so they become familiar with the wording and how to think through these types of tasks.

What item types are included in the Online PARCC?

The question types in Math are:

1. Drag and Drop
2. Drop Down
3. Essay Response
4. Extended Constructed Response
5. Hot Text Select and Drag
6. Hot Text Selective Highlight
7. Matching Table In-line
8. Matching Table Single Reponse
9. Multiple Choice – Single Correct Response, radial buttons
10. Multiple Choice – Multiple Response, check boxes
11. Numeric Response
12. Short Text
13. Table Fill-in

© Lumos Information Services 2018 | LumosLearning.com

Lumos StepUp® Mobile App FAQ For Students

What is the Lumos StepUp® App?

It is a FREE application you can download onto your Android smart phones, tablets, iPhones, and iPads.

What are the Benefits of the StepUp® App?

This mobile application gives convenient access to Practice Tests, Common Core State Standards, Online Workbooks, and learning resources through your smart phone and tablet computers.

- Eleven Technology enhanced question types in both MATH and ELA
- Sample questions for Arithmetic drills
- Standard specific sample questions
- Instant access to the Common Core State Standards
- Jokes and cartoons to make learning fun!

Do I Need the StepUp® App to Access Online Workbooks?

No, you can access Lumos StepUp® Online Workbooks through a personal computer. The StepUp® app simply enhances your learning experience and allows you to conveniently access StepUp® Online Workbooks and additional resources through your smart phone or tablet.

How can I Download the App?

Visit **lumoslearning.com/a/stepup-app** using your smart phone or tablet and follow the instructions to download the app.

QR Code for Smart Phone Or Tablet Users

Lumos SchoolUp™ Mobile App FAQ For Parents and Teachers

What is the Lumos SchoolUp™ App?

It is a free app that teachers can use to easily access real-time student activity information as well as assign learning resources to students. Parents can also use it to easily access school-related information such as homework assigned by teachers and PTA meetings. It can be downloaded onto smart phones and tablets from popular App Stores.

What are the Benefits of the Lumos SchoolUp™ App?

It provides convenient access to

- Standards aligned learning resources for your students
- An Easy to use Dashboard
- Student progress reports
- Active and inactive students in your classroom
- Professional development information
- Educational Blogs

How can I Download the App?

Visit **lumoslearning.com/a/schoolup-app** using your smartphone or tablet and follow the instructions provided to download the App. Alternatively, scan the QR Code provided below using your smartphone or tablet computer.

QR Code for Smart Phone Or Tablet Users

© Lumos Information Services 2018 | LumosLearning.com

Progress Chart

Standard	Lesson	Page No.	Practice		Mastered	Re-practice /Reteach
CCSS			Date	Score		
3.OA.A.1	Understanding Multiplication	9				
3.OA.A.2	Understanding Division	15				
3.OA.A.3	Applying Multiplication & Division	19				
3.OA.A.4	Finding Unknown Values	23				
3.OA.B.5	Multiplication & Division Properties	27				
3.OA.B.6	Relating Multiplication & Division	31				
3.OA.C.7	Multiplication & Division Facts	36				
3.OA.D.8	Two-Step Problems	43				
3.OA.D.9	Number Patterns	47				
3.NBT.A.1	Rounding Numbers	66				
3.NBT.A.2	Addition & Subtraction	70				
3.NBT.A.3	Multiplying Multiples of 10	74				
3.NF.A.1	Fractions of a Whole	85				
3.NF.A.2	Fractions on the Number Line	91				
3.NF.A.3	Comparing Fractions	97				
3.MD.A.1	Telling Time	108				
3.MD.A.1	Elapsed Time	113				
3.MD.A.2	Liquid Volume & Mass	118				
3.MD.B.3	Graphs	122				
3.MD.B.4	Measuring Length	130				
3.MD.C.6	Area	135				
3.MD.C.7	Relating Area to Addition & Multiplication	141				
3.MD.D.8	Perimeter	146				
3.G.A.1	2-Dimensional Shapes	169				
3.G.A.2	Shape Partitions	173				

Available

- At Leading book stores
- Online www.LumosLearning.com

Made in the USA
Lexington, KY
28 January 2018